Design & Build Electronic Power Supplies

Design & Build Electronic Power Supplies

Irving M. Gottlieb

TAB BOOKS
Blue Ridge Summit, PA

FIRST EDITION
FIRST PRINTING

Library of Congress Cataloging-in-Publication Data

Gottlieb, Irving M.
Design & build electronic power supplies / by Irving M. Gottlieb.
p. cm.
Includes index.
ISBN 0-8306-8540-5 ISBN 0-8306-6540-4 (pbk.)
1. Power electronics. 2. Electronic apparatus and appliances--Power supply—Design and construction. I. Title. II. Title: Design and build electronic power supplies.
TK7868.P6G64 1991
621.31′7—dc20 91-11243
CIP

TAB Books offers software for sale. For information and a catalog, please contact TAB Software Department, Blue Ridge Summit, PA 17294-0850.

Acquisitions Editor: Roland S. Phelps
Book Editor: Eileen Baylus
Production: Katherine G. Brown
Book Design: Jaclyn J. Boone
Cover photograph: Susan Riley, Harrisonburg, VA.

Contents

Preface

ALTHOUGH POWER-SUPPLY TECHNOLOGY HAS NEVER CEASED TO EVOLVE, significant changes in the overall art have not been an everyday occurrence. It is easy to see why. The mere availability of an improved component does not necessarily yield an improvement of the same magnitude in the performance of the overall power supply; as with the chain that is no stronger than its weakest link, it is usually the case that just one of a supply's many components generally does not make last year's model obsolete.

It is not economical for a manufacturer to have frequent changes in its production line, test, evaluation procedures, or its sales pitch. It is, indeed, a fact of life that there tends to be greater inertia against changing power-supply systems than you encounter in other equipment, such as computers, or in communications.

It isn't often that new circuitries or design approaches compel attention in power-supply equipment as is the case in other electronic technologies.

Having said this, however, it appears at this writing that meaningful change is just around the corner. After a long term of basically standardized techniques, the designer now finds it expedient to acknowledge and assess radically new power devices, control integrated circuits (ICs), and circuit topographies. Nor is it a matter of change for change's sake—the new approaches promise greater operating efficiency, more watts per cubic inch, better power line power factor, reduced electromagnetic and radio frequency interference, extended reliability, and decreased production costs. Interestingly, some long-known, but little-used design approaches now merit reconsideration.

The purpose of this book is to supplement and update material covered in other books covering power supply topics. But, rather than focus on "how to build" projects, the theme of this book will be to provide guidance and insights for the benefit of those who are technically knowledgeable, and in a position to profit from experimentation and creative engineering in a changing field.

Introduction

THE SUCCESS AND POPULARITY OF THE SWITCHING-TYPE POWER SUPPLY derives from an elegant premise, the accomplishment of regulation by simply turning a power device on and off for durations appropriate to maintaining a preset output voltage or current. Thus, instead of using the power device as a dissipative rheostat, the device is either on or off, and, ideally, never in a conductive state between these two extreme conditions. Such operation theoretically incurs no dissipative losses, tending therefore toward high efficiency, and low heat generation in the overall power supply.

Although such ideal switching performance cannot be realized in actual practice, the losses proved to be acceptably low in the popular 20-kHz designs. The switching technique in conjunction with 20-kHz operation enabled great reduction in the physical size of magnetics and filter components as compared to 60-Hz circuits. Moreover, the noisy operation of designs based upon audible switching rates was avoided. Such was the state of the art for a long time; it was satisfactory and individuals ran into considerable expense in attempts to appreciably increase the switching rate.

Nonetheless, it had been realized all along that higher switching rates could bring about further reductions in size and weight, possibly in cost, and conceivably in efficiency. Better power devices enabled conventional designs to be extended to the 30-kHz to 75-kHz region, but it became evident that you could all too readily get into an area of diminishing returns. Problems of cost, reliability, stability, and reproducibility on the production line plagued such supplies. Clearly, speedier power switches alone were not likely to extend switching rates into the desirable 100-kHz to several MHz region.

As might be anticipated, the solution to this dilemma appears to be via switching techniques other than the long-used pulse-width modulation. By duly recognizing the nonideal nature of conventional high-frequency switching, designers and experimenters have demonstrated different approaches yielding very high switching rates with all of the expected benefits, but with minimal dissipation in the power switch itself. To the designer and user, constrained by long-endured limitations, the new power supply technology is suggestive of having your cake and eating it too.

The ensuing pages deal with a wide variety of topics relevant to the design, improvement, and modernization of power supplies; interestingly, it will be found that appropriate blends of old *and* new ideas often lead to optimum results.

1

A look at the newer circuit techniques

IT HAS LONG BEEN REALIZED THAT MANY BENEFITS COULD BE EXPECTED from inverters and switch-mode power supplies as the switching rate is increased. Included in the menu of improvements would be size and weight reductions of magnetics and of capacitors, faster transient response, easier handling of radio frequency interference (RFI) and electromagnetic interference (EMI) problems, and lower manufacturing costs. This being the case, simple scaling techniques went a long way in fueling the evolution of design from 60 Hz to 400 Hz, and then to the several-kHz range. Aided by the transition from germanium to silicon power transistors, the 20-kHz technology came into prominence.

Switching at, or in the vicinity of, 20 kHz yielded immense benefits contrasted to the erstwhile 60-Hz equipment. This being the case, it was only natural to try for yet higher rates. A large measure of success accompanied 30-kHz and 40-kHz designs, but it was clear that a region of diminishing returns had been encountered. Even if exotic and costly switching devices were used, the poor high-frequency performance of electrolytic capacitors, transformer and inductor core materials, and of various diodes, ate heavily into the high theoretical efficiencies that were targeted. Thus, a development plateau was reached; higher ascent into the multikilohertz and into the megahertz switching range would have to await improved development of the circuit components as well as the switching device itself.

This is why the 20-kHz designs endured so long. However, with the advent of the power MOSFET switching device, component improvement received renewed impetus to advantageously use the higher switching rates inherent in this solid-state switch. This resulted in economic and reliable switching circuits in the vincinity of 100 kHz, and switchers utilizing rates of several hundred kHz have proven feasible.

At the same time, circuit techniques also have been improved to the extent that operation in the one- to several-MHz region can be attained by circumventing operating features long used at lower frequencies.

The invasion of the higher frequencies has not been the only aspect of change in power supply technology. New power semiconductors have emerged that can greatly extend the voltage, current, and power capabilities of 20-kHz designs. As you shall see, complacency with the state of the art is a short-lived luxury in the field of solid-state power conversion.

Poor performance of switching circuits at high frequencies

Technology has advanced to the point in which insufficient high-frequency response of the amplifying device is not necessarily the limiting factor to high-efficiency performance. For example, all devices suffer appreciable dissipative losses whether used in class A, B, or C modes of power boosting. The reason for this is such circuits permit simultaneous current through the device and voltage across the device.

For a long time, the solution to this obstacle was to develop switching and amplifying devices capable of extended high-frequency response. Such devices could maintain square waves of voltage and current, thereby keeping dissipation from simultaneous voltage and current at minimal levels. This approach is straightforward and cannot be faulted in concept. At higher frequencies, it turns out that other approaches tend to be more practical and less costly.

Consider, for example, a device that at some high frequency can no longer maintain square wave switching waveforms. Whether this comes about via transit time phenomena, charge-storage effects, or stray reactances, the result of the nonideal waveforms is invariably simultaneous current through the device, and voltage across the device for a nonacceptable percentage of the switching duty cycle. Fortunately, there are ways to prevent such dissipation. Although the high-frequency response of the device might seem to be the culprit standing in the way of high-efficiency operation, modification of the circuit operation can restore much of the efficiency lost in conventional switch-mode operation when higher switching rates are used.

How can this be accomplished? The common denominator of the several circuit modifications is that they prevent simultaneous current and voltage in the device. This can be done by altering the shape of the voltage and/or the current waveforms, by shifting their times of occurence, or by some commutation technique where current or voltage is made to be zero while the other is at a high level. Some of the known methods will be investigated. The application of such methods to power supplies might require a good measure of ingenuity because of the state of the art. On the other hand, megahertz switching supplies using such methods already have been marketed, and are certainly the progenitors of the compact, high-efficiency, low-cost power supplies destined to become commonplace.

Ideal switching devices

Manufacturers of high switching rate regulated power supplies tend to be very proprietary minded and have resorted to obtaining patents to prevent dissemination of their circuit techniques. (By high switching rate, you deal with the 100-kHz to several-MHz region; switchers below 100 kHz tend to be scaled extensions of popular 20-kHz techniques.) You can hardly fault these manufacturers when it is realized that power densities on the order of 30 watts per cubic inch have been attained, contrasted to about two or three watts per cubic inch in 20-kHz designs.

Even though published circuitries of the new art are difficult to come by, it is not hard to ferret out the basic philosophy of such high switching rates. Once having done this, the clever experimenter or the creative designer is in a good position to extract circuit building blocks from other technologies and combine these to yield the desired results. First it would be advantageous to review the true, as opposed to the ideal, nature of power switches as commonly employed in 20-kHz and somewhat higher frequency regulated power supplies. Figure 1-1 depicts the waveforms of an ideal switching device.

A usually encountered explanation of the benefits of pulse wave modulation (PWM) switching as contrasted to *linear* regulators is that the the power switch is either ON or OFF, and accordingly develops zero 1^2R losses. In the real world, this is not strictly so. As switching rates beyond 20 kHz are considered, the greater the departure from the concept of the ideal switch is. You'll find that the power device cannot be instantaneously changed from one conductive state to the other; there are definite rise and fall times associated with the switching process. Unfortunately, as the switching rate is increased, these rise and fall times constitute an ever greater percentage of the total on time. Let's see what is so detrimental about this.

It might initially appear that the times of rise and fall would just transform the square switching wave into a trapezoidal waveform, and that this might be somehow compensated for by extending the duration of on time. However, the time spent by the power switch between fully on and completely off is dissipative because the switch has voltage impressed across it while current is passing through it. It is behaving essentially as a rheostat, not as an ideal switch. Thus, the higher the frequency of switching, the greater these so-called *switching losses* are. Here you can see one reason why 20 kHz long loomed up as a practical compromise between switching rate and efficiency.

Real-world devices and their trapezoidal switching waveforms

In bipolar transistors, it is the storage of minority carriers that is primarily responsible for the inability to switch conductive states instantaneously. It is as if these devices had internal capacitors in which charging and discharging involved definite time constants. This phenomenon is absent in power MOSFETs, these being majority-carrier devices. Thus, such MOSFETs can pro-

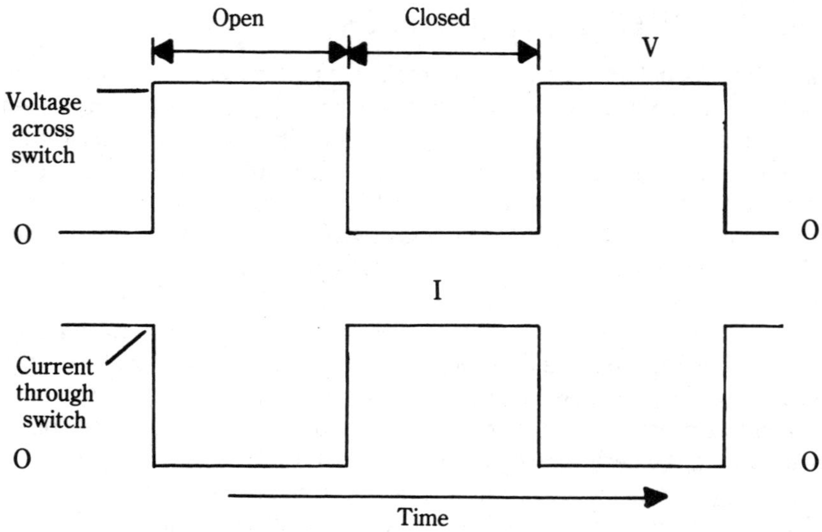

1-1 Voltage and current waveforms accompanying an ideal switch.

vide much greater switching rates than power bipolar transistors. Because of stray reactances in the packaging and in the circuit of power MOSFETs, these devices also will incur dissipation in the rise and fall of the switching waveform if the switching rate is made too high. Several hundred kHz rates have proven practical. At low switching rates, the waveforms of practical devices tend to resemble those of Fig. 1-1. At higher rates the waveforms of all actual devices deteriorate into the trapezoidal pattern shown in Fig. 1-2.

Both bipolar and MOSFET switching devices exhibit another loss that prevents ideal switching operation. Both show I^2R loss during the on conductive state inasmuch as neither is able to develop zero resistance. The bipolar transistor is the better performer here; its saturation voltage tends to be lower than is manifested in a similarly rated power MOSFET. Moreover, the loss due to on resistance in the power MOSFET becomes progressively worse with higher voltage ratings. With bipolar transistors, the voltage drop across the device remains for practical purposes negligible at higher voltages. In fairness, however, it must be pointed out that the early power MOSFETs were responsible for much of their bad reputation in this regard; there has been dramatic improvement in the lowering of the on resistance of power MOSFETs. On resistance dissipation is also shown in Fig. 1-2.

The on resistance of both devices is substantially constant with frequency, yet can indirectly lower the allowable switching rate. This is because of the thermal consequences of all of the losses in a switching device. A point will be reached where attempts at a higher switching rate will be unwise because of the temperature rise from the sum of the losses within the device.

The inability to maintain square or rectangular waves of voltage across, and current through the power switch is the prime cause of deleterious switching

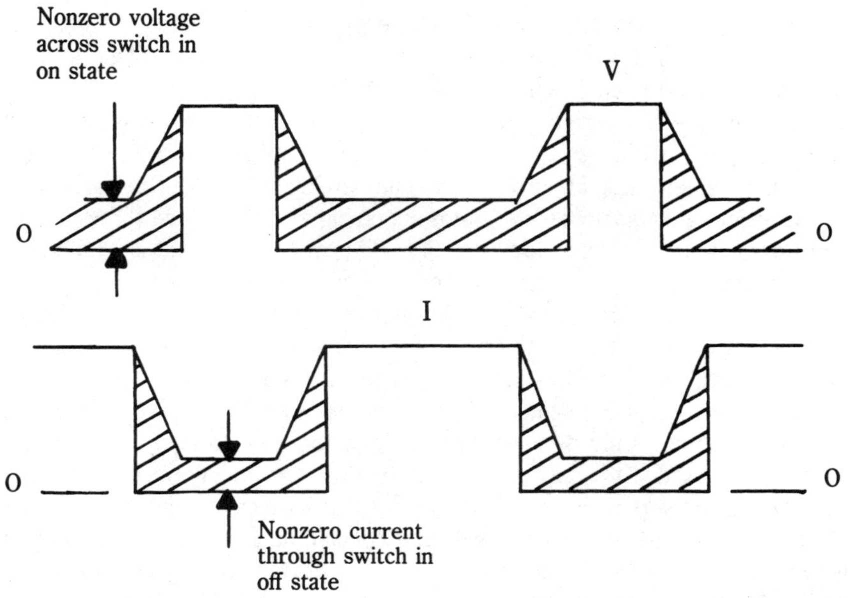

1-2 Voltage and current waveforms in actual switching devices.

losses at higher switching rates. Trapezoidal switching waveforms bring back the dissipative rheostat action that the transition from linear to switch-mode regulation was supposed to eliminate. That being the case, our quest for higher switching rates must center on producing as good a square wave as possible, or alternatively by using a different switching technique than that in conventional 20-kHz regulating supplies.

Sine wave switching reconsidered

The use of sine waves rather than square waves in switching power supply, inverter, and regulator circuits is far from new. Until now, the chief reason for turning a power device off and on with sine waves has been the reduction of RFI and EMI. This follows from the fact that a pure sine wave contains no harmonic energy. The major drawback of many of these sine-wave circuits was low operating efficiency with attendant heat-removal problems. However, it is not necessarily true that low efficiency is inherent in the use of sine waves. Conversely, the appropriate use of sine waves greatly reduces the principal loss in high-frequency square-wave circuits, simultaneous voltage and current at the switching device.

Unfortunately, the idea that loss of energy is associated with sine waves per se is quite deeply entrenched. At its roots, the argument against the use of sine waves claims that the transformation of a square wave into a sine wave entails the removal of all of the harmonic energy in the square wave except the fundamental, or first harmonic. That being the case, you must reconcile to the "loss" of a considerable portion of the energy initially contained in the original square wave. This line of reasoning reinforces its assertion by calling to attention that the "stripped-down" sine wave obviously comprises less area when drawn alongside the parent square wave; less area infers less energy, so the argument goes. Although this claim seems to be imbued with a good deal of "common sense," it is important that it be debunked as not necessarily so.

Consider a square-wave generator supplying a resistive load through a perfect (lossless) band-pass or low-pass filter. The voltage across the load and the current through it will be sinusoidal, and a certain amount of power will be delivered to the load. This simple situation is depicted in Fig. 1-3. If the filter is removed, you know that additional power will be delivered to the load, now impressed with square-wave voltage and current. The interpretation of this is important. Although the sinusoidal power is less than the square-wave power, the "lost" power is not dissipated en route to the load. Indeed, it is not dissipated or lost at all—it simply is not utilized. Notwithstanding the inequalities of the two load powers, it is true that virtually 100% efficiency prevails for either square wave or sine wave load power. In particular, no inefficiency or power loss is incurred because of the sine-wave voltage and current in the load. Do not look under the rug for second-order effects in this simple setup, assuming the generator has a very low output impedance, and that a low frequency is used should dispel the temptation to find side effects in the experiment.

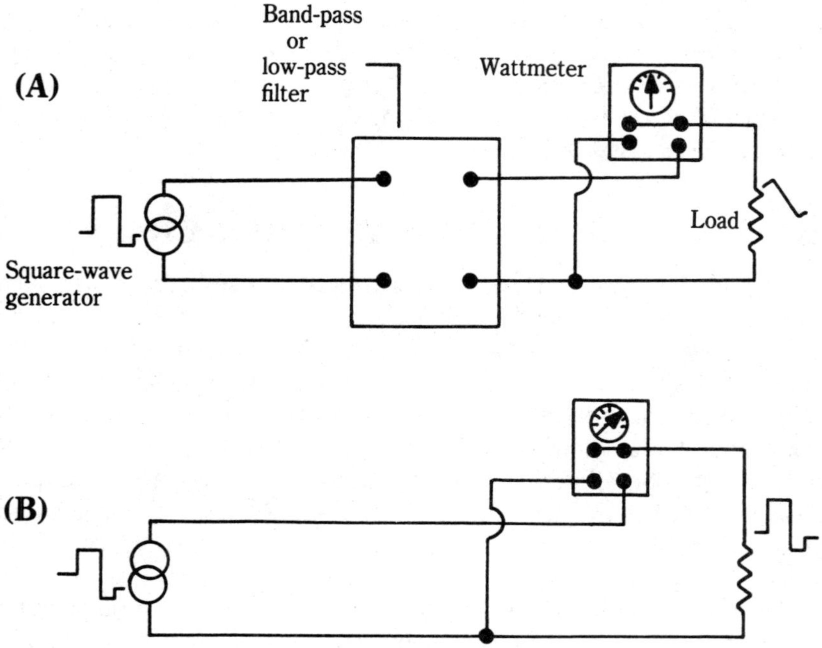

1-3 Square wave load power exceeds that of its sine wave fundamental frequency.

Applying square waves to a sine wave rated load

Now that it is established that dissipative losses are not necessarily produced by empowering a load with a sine wave derived from a square wave source, it is important to consider the converse situation and its practical ramifications. Now deal with a fairly common dilemma—a solid-state inverter or power supply with an AC square-wave output voltage, and a motor specified by its manufacturer for operation from a sine-wave source. The not so easy to answer question is, what should be the rms voltage of the square wave delivered to the motor?

A quick answer in a qualitative sense is that the motor should require less square-wave voltage than its rated sinusoidal voltage for equivalent motor performance. The rationale for this judgement usually is based on the notion that there is more energy content in the square wave than in a sine wave of the same rms value. Although the answer derived from this false logic might sometimes be correct, it should be recognized that other cause-and-effect relationships are responsible, not the intuitive logic employed.

Much will depend on the size and type of motor, and the frequency of its square-wave power source. Often the self-inductance of the motor will be sufficient to effectively reject harmonic currents, thereby allowing the sine wave fundamental frequency to produce the actual electromechanical torque. Recall that current is the parameter that is most basic in motor action. At the same time, it remains true that current is proportional to impressed voltage. As long as you are consistent, you can deal with either parameter. It happens that it is somewhat more relevant to the practical situation to consider voltages; this is especially true because you wish to know what square-wave voltage to apply to your sine-wave-rated motor.

First, investigate the consequence of the filtering produced by the motor inductance on the actual applied motor voltage. The square wave of voltage originally comprises the fundamental frequency (the first harmonic) and a large number of higher odd harmonics, that is, the third, the fifth, the seventh, and so on. If you consider the harmonic rejection as because of an ideal filter, all that will be left will be the sine wave fundamental frequency. Is the rms value of this sine wave (which the motor "sees") greater or less than the rms value of the original square wave? Rather than rely on a snap judgement, it will be better to see what the mathematics of waves has to say. You can, however, make use of Fig. 1-4 to understand the situation.

First, review the important definitions pertaining to the sine wave. For convenience, these definitions relate to the peak value, a wave property that is obvious from inspection. If you let the peak value of the sine wave be represented by V_P then the rms value = 0.707 (V_P). Thus, the rms value of a sine wave with a peak value of 100 volts is 70.7 volts. The reciprocal relationship is also useful: Because 1/0.707 = 1.414, it follows that the peak value of a sine wave is derived by multiplying its rms value by 1.414. It will be recognized that

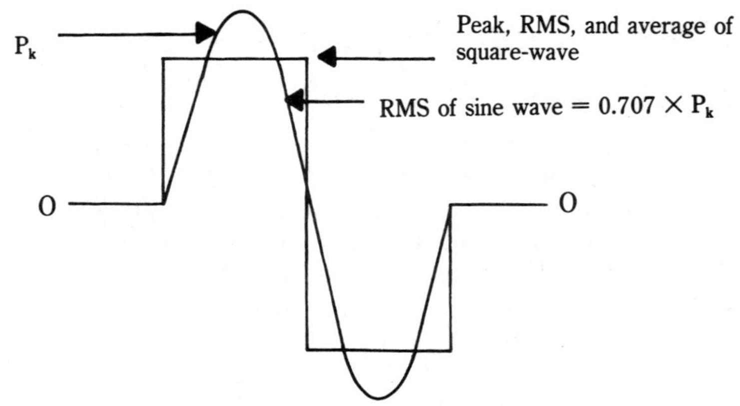

1-4 Relationship between square wave and its fundamental-frequency sine wave.

1.414 is the square root of 2. However, in this book you will not deal with fundamental derivations; these and their mathematical underpinnings can be found in texts devoted to basic relationships in electricity. Note that the rms or effective value of the sine wave is not the same as its average value, which is given by 0.636 (V_P).

So much for the sine wave. Now, the interesting thing about the square wave is that its peak value, its rms value, and its average value are all one and the same. Keeping these matters in mind, it is helpful to contemplate the waves depicted in Fig. 1-4. What you see here is a square wave together with its fundamental sine wave. (Assume all the higher harmonics than the first have been rejected by an ideal filter.) The strange thing about these waves is that the residual sine wave has a higher peak voltage than does the "parent" square wave. Your initial reaction to this almost has to be that it can't be so.

That it can be so is readily shown by drawing in several of the rejected harmonics. This is done in Fig. 1-5. You now can see that if the square wave were reconstituted by restoring a great number of harmonics, the protruding peak of the fundamental frequency seen in Fig. 1-4 would disappear. So, in a qualitive way, it can be accepted that the filtering action of motor inductance can leave a fundamental frequency sine wave with a peak voltage exceeding that of the square wave.

Quantitatively, it can be shown that the peak of the sine wave is $4/\pi$, or 1.27 times the peak value of the square wave. If you multiply 0.707 by 1.27, you find that the rms value of the fundamental frequency sine wave is 0.898 times the amplitude of the square wave. So, notwithstanding its high peak value, the rms value of the sine wave is less than the amplitude of the square wave. These mathematical relationships imply that the motor should be impressed with a square wave of 1/0.898 or 1.11 times the rated rms sine-wave voltage.

The situation you have worked out tends to be a paradox to those who haven't given proper consideration to the relationship between the impressed square wave and the sine wave "seen" by the motor. At the same time, the

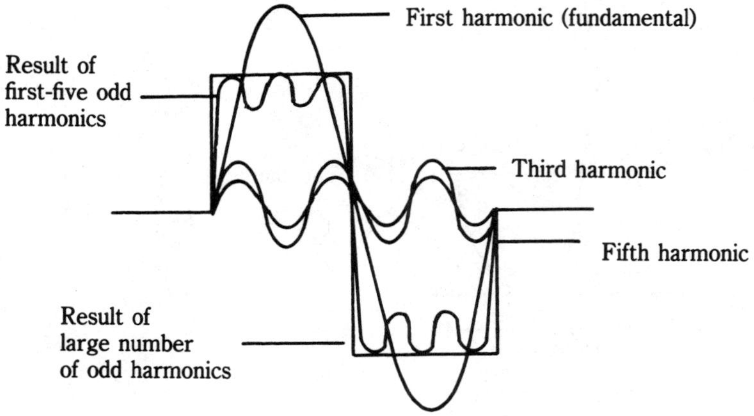

1-5 More-detailed view of Fourier wave composition of the square wave.

mathematical treatment was based upon ideal circumstances that might or might not be approximated by actual motor systems. Indeed, there might be sufficient harmonic energy in the motor to roughen its torque characteristics, and to contribute to the temperature rise from eddy currents and hysteresis. The overall result of these nonideal situations is that the square-wave voltage applied to the motor might, indeed, have to be less than the rated rms sine-wave voltage. If so, however, it is a good idea to understand the practical compromises that have been made with "pure" theory.

Sometimes you will find that the square-wave voltage indicated by theory can be successfully used if the rudimentary filter is inserted between the square wave source and the motor. This is usually an L section making use of a practically sized inductor and capacitor. Such a filter makes it unnecessary to depend entirely on motor inductance for attenuation of the harmonic energy. The actual LC values can be determined empirically, the objective being to provide minimal attenuation of the fundamental frequency, low dissipative losses, and an acceptable compromise of physical size, weight, and cost. The fact that the lowest harmonic you wish to attenuate is the third, rather than the second harmonic, makes it relatively easy to experimentally produce a satisfactory filter. If the filter does a reasonably good job of harmonic rejection, the motor will, indeed, deliver expected performance when impressed with a square wave with the theoretical 110% of the rated sine-wave voltage. Because of size and cost of the LC filter elements, it is usually more practical to use filters with small subfractional horsepower motors than with larger motors.

The sine wave with a DC bias

Another situation sometimes encountered with sine waves in power-supply circuitry is shown in Fig. 1-6. Here, it is seen that a sine wave with a 10-volt peak-to-peak amplitude is superimposed on a five-volt DC level. The need to know the rms value of these two "waves" acting in a linear circuit might arise. Con-

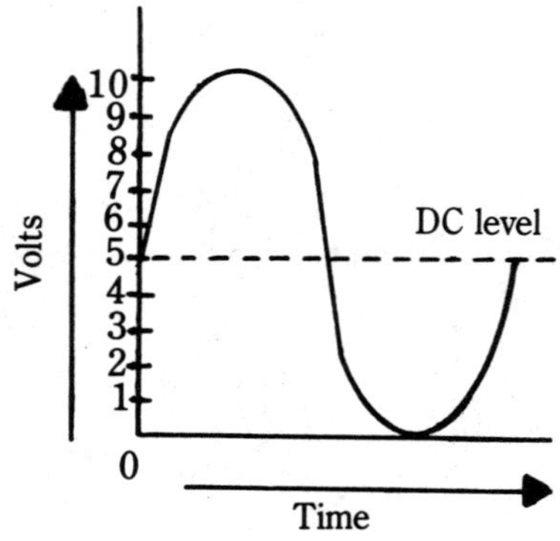

1-6 The sine wave together with a DC level.

siderable confusion often attends the pursuit of the solution to problems of this kind. Although it is true that analytical geometry and calculus can be used to work out the answer from basic principles, a much simpler calculation can be employed to yield a quick solution.

First, it should be recognized that from a mathematical viewpoint, there are, indeed, two frequencies involved. One is the frequency of the sine wave, the other is the "frequency" of the DC level, which is zero. Other than this, you need not concern yourself further with frequency; the basic idea is that there is a resultant rms value without regard to the actual value of the two frequencies; there is nothing unique about one of the frequencies being zero.

Once over this hurdle, the sought rms value derives from the following procedure:

1. Obtain the rms value of each of the components.
   For the sine wave, the rms value is $0.707 \times 5 = 3.535$ volts
   For the DC level, the rms value $= 5$ volts.
2. Calculate the sum of the squares of these rms values, thus

   $(3.535)^2 + (5)^2$
   $= 12.5 + 25$
   $= 37.5$

3. Obtain the square root of the sum of the squares

$$\sqrt{37.5} = 6.124 \text{ volts}$$

This is the value that would be indicated by a true rms voltmeter. However, a caution is in order because not all so-called "true" rms meters actually respond to the presence of DC. Note that 6.124 is less than the 8.535 volts obtained by simple addition of the two rms values—the common mistake made in handling this type of situation.

The class-F rf amplifier

A good place to start the investigation of alternate switching circuits is the class-F amplifier thus far used primarily in vhf and uhf work. This rf power amplifier is much more efficient than the long-used class-C amplifier. (Bear in mind that, although these circuits are referred to as *amplifiers* they actually operate as *switches*; there lies the relevancy to switch-mode regulators used in power supplies.) The major deterrent to higher efficiency in class-C amplifiers is the fact that there is simultaneous voltage and current over part of the switching cycle. This shortcoming is inherent in class-C operation even though the tube, MOSFET, bipolar transistor, or other device exhibits more than adequate high-frequency response.

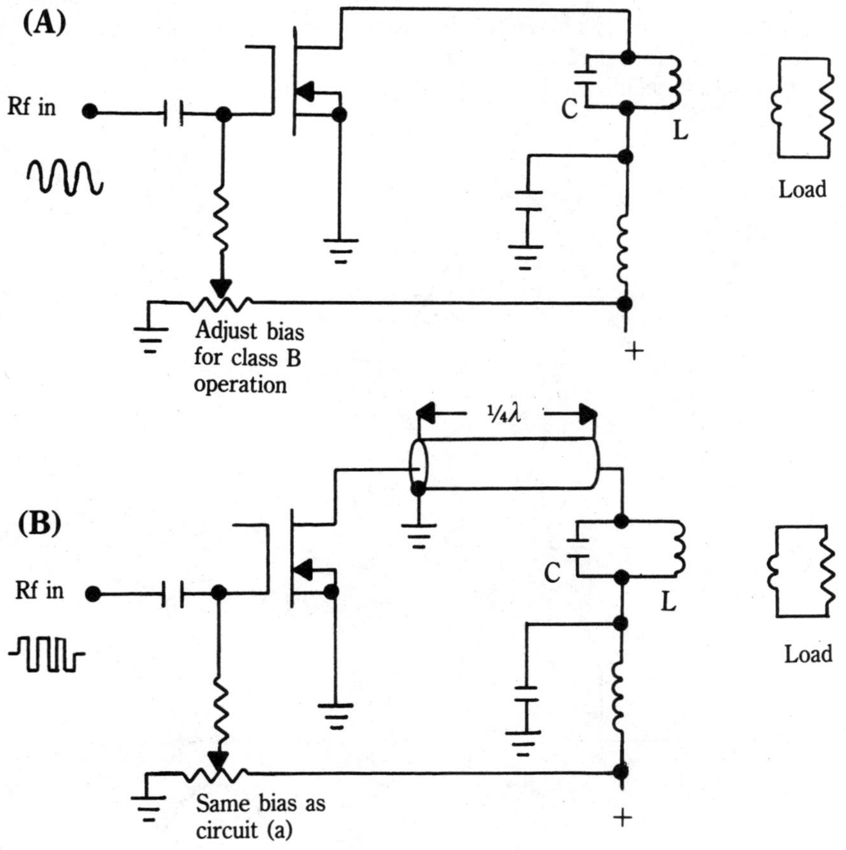

1-7 Class-B and class-F power MOSFET amplifiers.

Topographically, the class-F amplifier resembles the class-C amplifier, but with a quarter-wave transmission line inserted between the device and the output tank circuit. The device is biased for approximate class-B operation. This could lead you to suspect that the efficiency would be even less than that achievable with class-C operation. However, this is where the quarter-wave transmission line asserts its influence. The line functions as a ***waveshaper***, converting the otherwise half sinusoid of plate, drain, or collector voltage into a square wave. With such a voltage wave, there is never simultaneous device voltage and device current. Inasmuch as the quarter wave line is not a nonlinear element, its waveshaping properties might not be self-evident. The class-B amplifier and the improved circuit, the class-F amplifier, are shown in Fig. 1-7.

Consider first the action of the parallel resonant output tank. This circuit supports the fundamental frequency, f_1, of the harmonic-rich wave impressed on it. At the same time, it short circuits all harmonics of f_1. Next, recall the unique behavior of quarter wave lines—a short circuit at one end becomes a high impedance (ideally infinite) at the other end; also, this behavior exists for not only f_1, but for all of the higher odd harmonics. That is, the impedance inverting action occurs for f_1, $3f_1$, $5f_1$, $7f_1$, and so on. In contrast, such a line

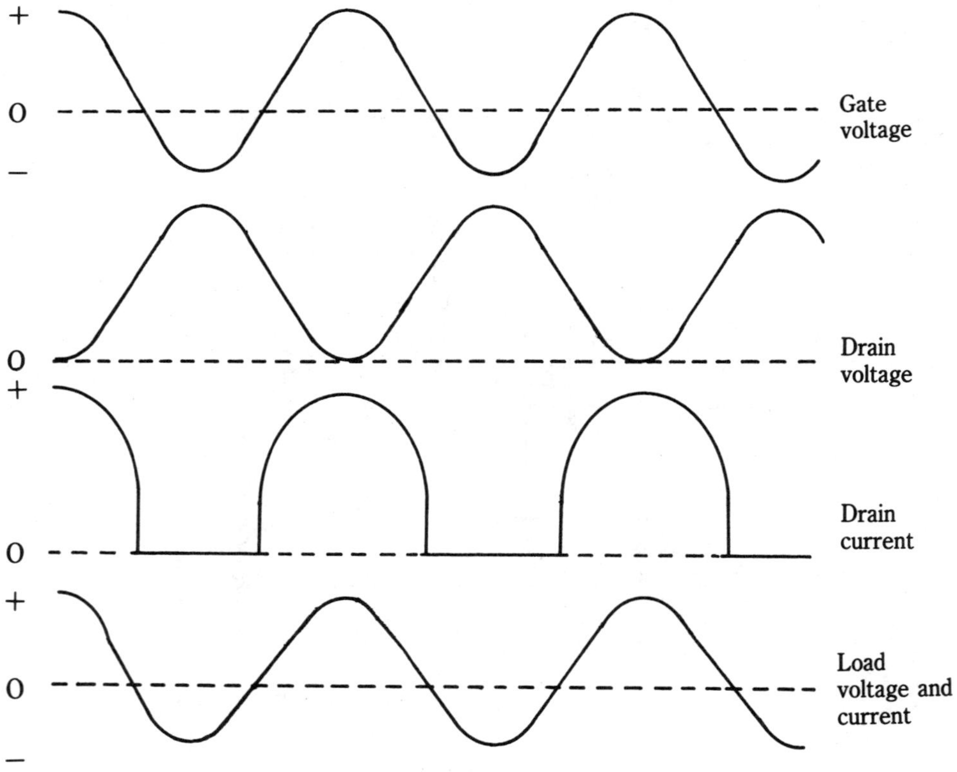

1-8 Waveshapes in the class-B amplifier.

simply repeats the impedance level it "sees" for even harmonics of f_1 such as $2f_1$, $4f_1$, $6f_1$, and so on.

Contemplation of the phenomena shows that the insertion of the line in the circuit does nothing to alter even-harmonic voltages at the output (drain) of the device—tank action simply repeats at the drain. It is as if the tank circuit was directly connected to the plate, drain, or collector of the device. With regard to odd harmonics, however, interesting things happen.

The first odd harmonic voltage is, of course, f_1 itself, the fundamental. At f_1 the high impedance "seen" at the parallel resonant tank circuit will be reflected as a relatively low impedance at the output electrode of the switching device. (In practice, this impedance-level can readily be made near optimum for efficient power transfer from device to load by appropriate selection of the characteristic impedance of the line, and of the operating Q of the tank circuit.) The important point here is that the fundamental frequency voltage f_1, is supported at the output electrode of the switching device. All higher odd harmonic voltages are also supported because the sending end impedance of the line is very high, as determined from the short circuit "seen" at the load end of the line (the LC tank circuit). Note the harmonic-rich square wave fed to the gate contrasted to the sine wave input for the ordinary class-B amplifier.

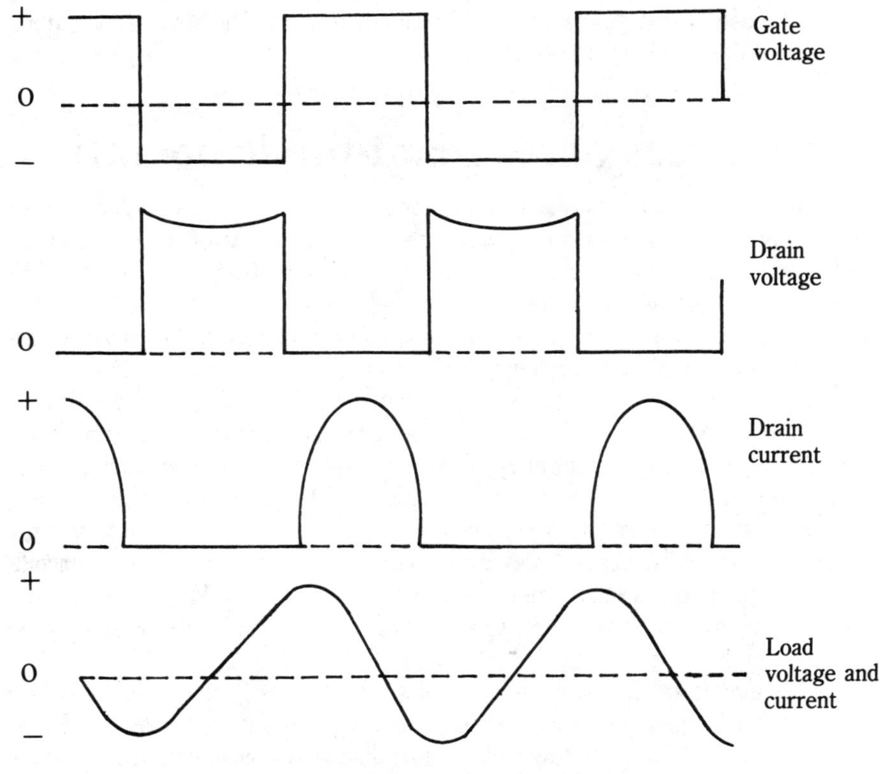

1-9 Waveshapes in the class-F amplifier.

From these actions, the following situation prevails: The voltage and current of the output tank is sinusoidal in normal fashion. At the output electrode of the switching device, there is a summation of harmonically related voltages comprising the fundamental frequency, f_1, and numerous odd harmonics of f_1. It fortuitively happens that the Fourier composite of these harmonic voltages is a near square wave. In effect, the quarter wave line has isolated the output electrode from the tank and load circuits for odd harmonics of f_1, but has nonetheless allowed normal operation of the tank circuit for f_1, itself.

The consequence of the waveshaping property of the line is evident from the waveform diagrams of Figs. 1-8 and 1-9. Whereas in ordinary class-B operation, there is simultaneous current and voltage in the switching device for a portion of the operating cycle, this no longer occurs in the class-F circuit containing the quarter wave line. An important source of power dissipation thereby is eliminated.

In practice, it is unwieldy to implement the class-F amplifier at low frequencies. But, how low is low? I have attained useful results in the vicinity of 10 MHz; it probably would be desirable to get down to several MHz for power supply applications. Using a power MOSFET and miniature coaxial line, nature tends to be on our side because the physical length of such a line (which is coiled as a solenoid) is only about 60% of its electrical length.

It is not inconceivable that future development of power supply technology will make use of switching rates considerably exceeding the several megahertz frontier of the present state of the art.

Advantages of resonant-mode operation

Resonant-mode operation of a switching power supply generally confers worthwhile advantages in both design and operation as compared to conventional pulse-width modulation commonly used in so-called "20-kHz switchers." (We are probably dealing with a range of switching rates from 20 kHz to about 100 kHz, whereas the resonant-mode supplies usually operate in the several hundred kHz to several MHz region.)

One of the notable features of resonant-mode operation is a worthwhile reduction of EMI and RFI. This comes about because the switching transitions occur at or near zero current or zero voltage. Also, part of the circuitry handles sine waves rather than harmonic-rich square or rectangular waves. Another fortuitous situation tending to keep electrical noise at bay is the effect of stray inductance and stray capacitance in resonant-mode operation. These unavoidable stray parameters tend to degrade clean switching in PWM supplies. In resonant-mode switchers, however, stray reactance often can be absorbed into the resonant tank circuit.

Although it might incur its own dissipative losses, the resonant-mode switcher tends to reduce the switching losses generally accompanying PWM circuits, especially at higher switching rates. The culprit is simultaneous voltage and current in the switch. This condition is neatly circumvented in the resonant-mode switcher because of the nature of the shock-excited waveform; zero current switching automatically occurs.

A nice thing about resonant-mode operation is that many conventional power supply circuits lend themselves well to conversion to resonant-mode operation. Two basic system changes are necessary. A resonant "tank" must be inserted at some point ahead of the output rectifier. A resonant mode or FM-type control IC must be substituted for the PWM control IC. Sometimes casual inspection does not suffice to identify a resonant-mode circuit. This is because at high frequencies, the leakage inductance of a transformer can substitute for a physical inductor, and the capacitance of semiconductor devices, together with wiring capacitances, can substitute for a physical capacitor. Proprietary designs often use such techniques, and marketing departments like to boast of such departures from "ordinary" practice used by the less creative competition.

Using sine waves for switching power supplies

We have covered a number of apparently independent topics dealing with sine waves because the newer high-frequency power supplies make use of sine wave circuitry. In a broad sense, all sine wave circuits will be found to have some relevancy to the techniques used in these power supplies. What has not yet been made clear is the manner in which sine wave technology can compete with the square wave switching circuits up to this time used in high-efficiency supplies. Inasmuch as it has been pointed out that square-wave techniques become inefficient at high frequencies because of degradation of the square wave into trapezoidal waves, it would seem that sine waves would represent the ultimate deterioration of the square wave. Thus, the use of sine waves for switching should be manifested by high losses—these arising from operation of the power switch in its linear region, that is, from simultaneous presence of current through the switch, and voltage across it.

It is indeed true that high dissipation would occur if you merely substituted a sine wave to control the switching behavior. Such a technique would be suggestive of a class-B amplifier. It actually has been used when the emphasis has been on low RFI and EMI. It deserves consideration for low power levels, where losses and heat removal are not major factors in design or operation. It is obvious, however, that this is not the appropriate approach to higher power levels. Yet, it is also tantalizing to contemplate the benefits of sine wave operation if the switching losses could be greatly reduced. From experiences with class-C and class-F rf amplifiers, and class-D audio amplifiers, it appears that operation at high power levels, high frequencies, and high efficiency should be achievable.

The published and proprietary sine wave power supplies might differ somewhat in circuit topography, but appear to have one thing in common—the use of shock excitation to produce the sine wave power. See Fig. 1-10. The compelling feature of shock excitation is that it enables the switching device to be square-wave controlled but to be spared the simultaneous action of current through it

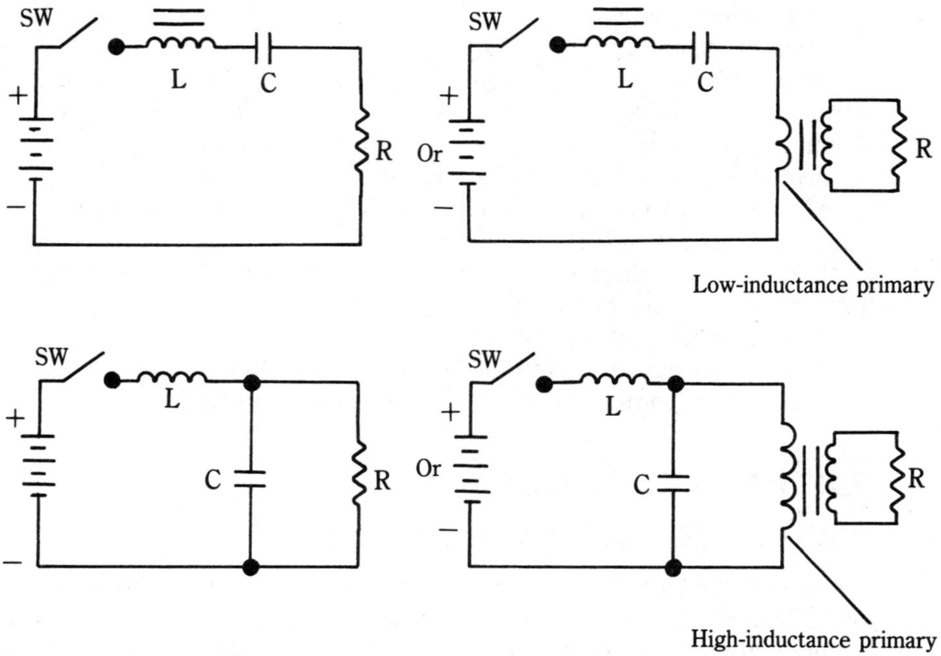

1-10 Basic switching circuits for resonant-mode supplies.

and voltage across it. Essentially, one has a switch associated with a resonant circuit. For regulation purposes, it is common to vary the frequency of the switching pulses instead of using the pulse-width modulation.

The secret of using sine waves for near zero switching losses is shown in Fig. 1-11. The switching waves illustrated are typical for resonant mode supplies. Note that the current wave and the voltage wave do not overlap. Thus the switching occurs at substantially zero current.

The resonant mode switcher

Most of the higher frequency inverters, converters, and regulated supplies generate sine or near sine waves as the result of LC circuits operating in the vicinity of resonance. Such techniques enable efficient operation in the 100-kHz to several-MHz segment of the spectrum. Other advantages accrue also, such as reduced EMI and RFI, more power capability in smaller physical volume, and enhanced cost-effectiveness. The latter is largely due to the reduced need for exotic power switching devices.

The fact that the power switch produces sinusoidal, rather than square wave current, damped enthusiasm at first for the new design approach. For engineers had expended much effort over a long period of time doing everything possible to upgrade the square wave switching performance of pulse width mod-

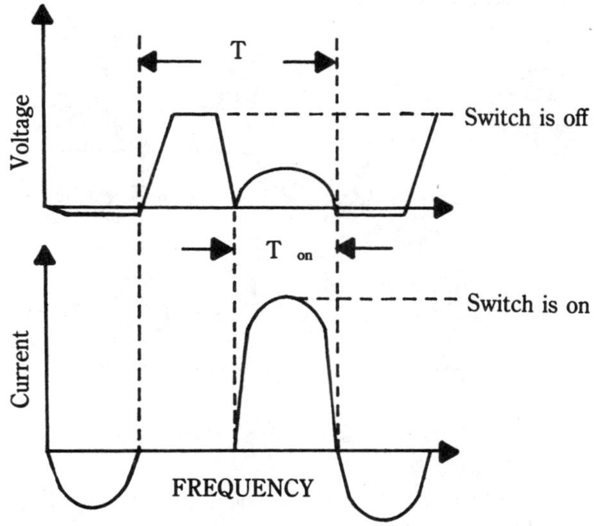

1-11 Secret of the sine-wave switch in resonant-mode supplies.

ulated power supplies. To fall back upon sine waves certainly appeared to be a step in the wrong direction. However, you need not resort to mathematical analysis or sophisticated circuit theory to see that it wasn't the nonsquare wave switching per se that previously led to high losses and low efficiency; it was the way in which the sine waves were used. The real basis for efficient switching is the avoidance of simultaneous voltage and current in the power switch during the operating cycle. If this can be done, sine wave switching current clearly exhibits the potential of high efficiency operation. This is readily seen by merely experimenting with voltage and current waves laid out on a sheet of graph paper. Hardware considerations need play no part in such a mental exercise.

To circumvent the possibility of simultaneous voltage across the power switch and current through it, you can visualize designs based upon zero current switching, zero voltage switching, combinations of these switchings, waveshaping techniques, and displacement of the timing of either the voltage or current wave. Rather than attempt to use pulse-width modulation for regulation purposes, most of these higher frequency switchers use frequency modulation. By varying the frequency of fixed on-time pulses applied to a resonant circuit, the output derived from the resonant circuit is electronically varied to achieve regulation. Thus, these resonant-mode supplies have VCO's and monostable pulse generators, rather than pulse-width modulators in their control electronics.

Why has resonant-mode operation arrived so late?

Resonant-mode operation of switching power supplies is actually an old concept and has been used in high-power SCR circuits for welding, ultrasonic, and other

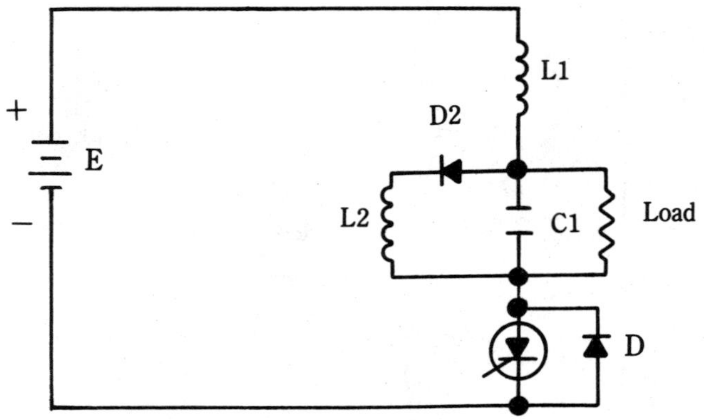

1-12 Single SCR sine-wave inverter operating in the resonant mode.

applications requiring a great deal of electrical energy. See Fig. 1-12. These power supplies seldom operate beyond the several tens of kHz region, and usually have been unregulated. Because of the resonant-mode operation, the keynote of these power supplies has been high efficiency. Also, reliable commutation of the SCRs made the design approach attractive. For many years, however, the technique did not gain popularity for systems ordinarily serviced by the millions of PWM 20-kHz switch mode supplies.

Because it was known that resonant-mode operation could, in principle, enable extension of switching rates to the megahertz region, and could greatly increase power-to-volume ratios, you must suspect the otherwise attractive idea was plagued by real or perceived disadvantages. This, indeed, had been the case. The story is not unlike the prelude to the eventual popularity of the PWM switcher.

As might be suspected, suitable switching devices and components had to await their turn in the course of technological evolution. And, even when such hardware became available, real progress could not be made until the advent of dedicated control integrated circuit (ICs). The stage is now set for advance from the 20-kHz region to switching rates of several-hundred kilohertz and higher. It is only natural, however, to ask whether resonant-mode supplies have any disadvantages.

The simple answer is that there are some adverse tendencies, but that these can be overcome readily. Resonant-mode switching circuits tend to produce peak switch currents some several times greater than their counterpart PWM circuits. Three things can be done to keep such peak currents manageable. The Q of the resonant circuit must not be too low—arrival at an optimum Q can be, admittedly, an empirical blend of art and science. The selection of switching devices with low conductive losses is very helpful in preventing degradation of efficiency. Finally, the dedicated ICs now becoming available often have provisions for both voltage-mode, and current-mode sensing. Current-mode sensing makes feasible cycle-by-cycle monitoring of peak current, and facilitates such protective functions as automatic overload shutdown, current limiting, soft start, and power factor correction. Whatever function is carried out, the change always takes place at the end of a switching cycle—never in the middle of a cycle, which would create dangerous transients and electrical noise. This current-sensing technique also improves line regulation and tends to relax feedback compensation problems.

Application of current-mode control to resonant supply

A common way of implementing current-mode control in resonant-mode supplies is shown in Fig. 1-13. Dedicated control ICs intended for resonant-mode circuits are often designed to use this technique of sensing the switching current. In Fig. 1-13, L1-C1 make up the series-resonant "tank." $T1_p$ is the primary of the output transformer; its inductance is sufficiently high so that its effect on the L1-C1 resonant frequency is not great. $T2_p$ is the primary of a current transformer and often makes just one turn. In this case, the inductance of $T2_p$ is sufficiently low so that its insertion in the series-resonant circuit has negligible effect on circuit behavior. The secondary of this current transformer then connects to appropriate terminals of the control IC.

The previously described situation provides one feedback loop in the power supply. A second feedback loop is also present—it is none other than the conventional voltage-mode sensing circuit. It operates in the ordinary way, that is, it causes a servoaction in such a manner that the output voltage is regulated. Keep in mind, however, that the control technique is not via pulse-width modulation. Rather, in the resonant-mode supply it is by variation of the constant-duration switching rate. Thus, it is an FM technique making use of the impedance curve of the resonant circuit.

Although you must grasp new ways of looking at circuit action, it can be appreciated that the resonant-mode supply is not more complex than the PWM type. In both cases, the appropriate control IC reduces the entire system to just a few building blocks.

Circuit topography of a resonant-mode power supply

The simplified schematic diagram of Fig. 1-14 conveys the general idea underlying many resonant-mode power supplies. Although there is a topological resemblance to the "conventional" switch-mode supply, it should be noted that the output transformer "sees" a sine wave, rather than a PWM waveform. This stems from the insertion of the series-resonant tank. This resonant circuit is shock excited by a variable rate rectangular wave in which ON time remains constant. Appropriately, circuits of this type often are called FM systems. Regulation occurs as the consequence of working on the impedance slope of the resonant circuit.

For successful implementation of this technique, it is desirable that dedicated FM control ICs be used; this is symbolized by the dashed enclosure of the

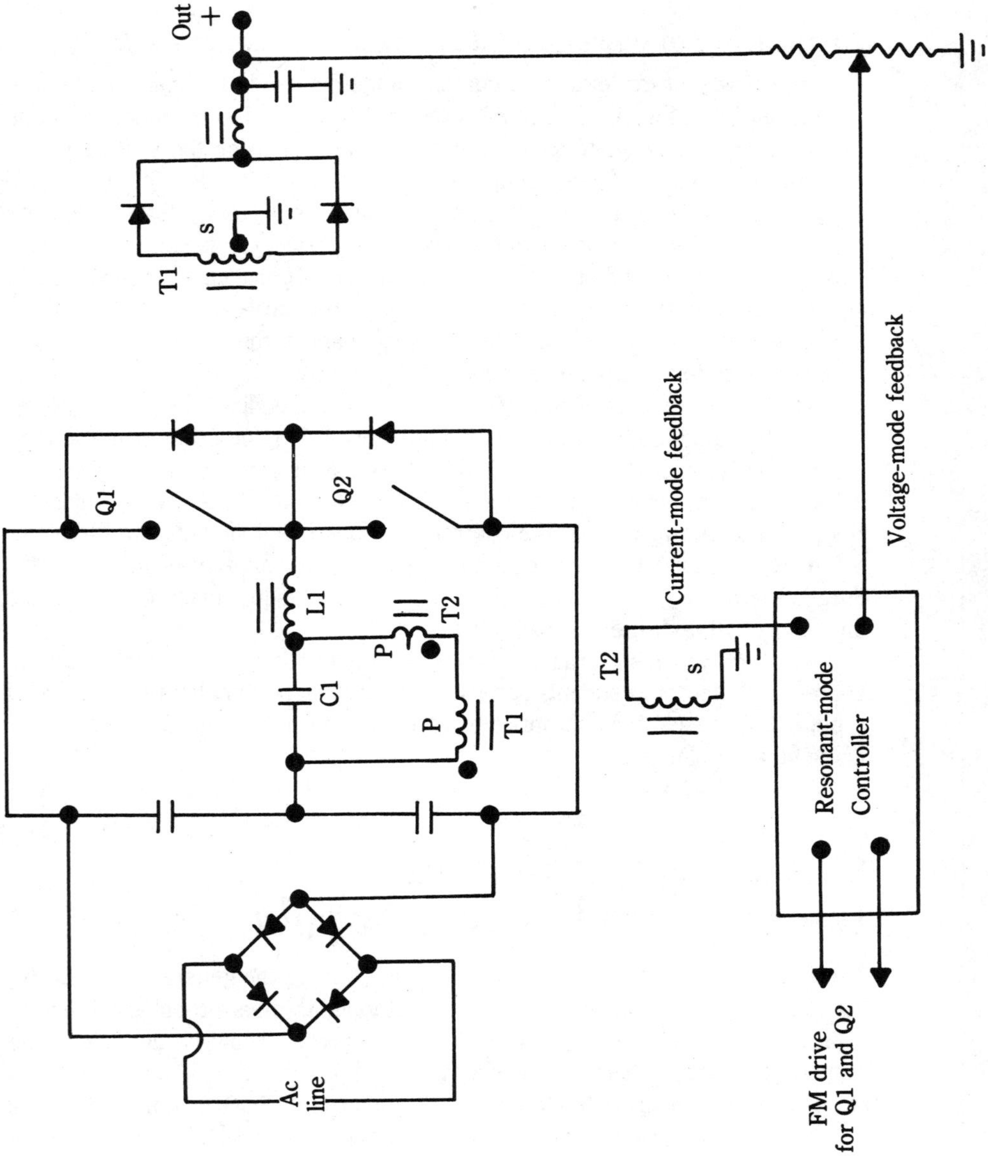

1-13 Typical layout for a half-bridge resonant-mode power supply.

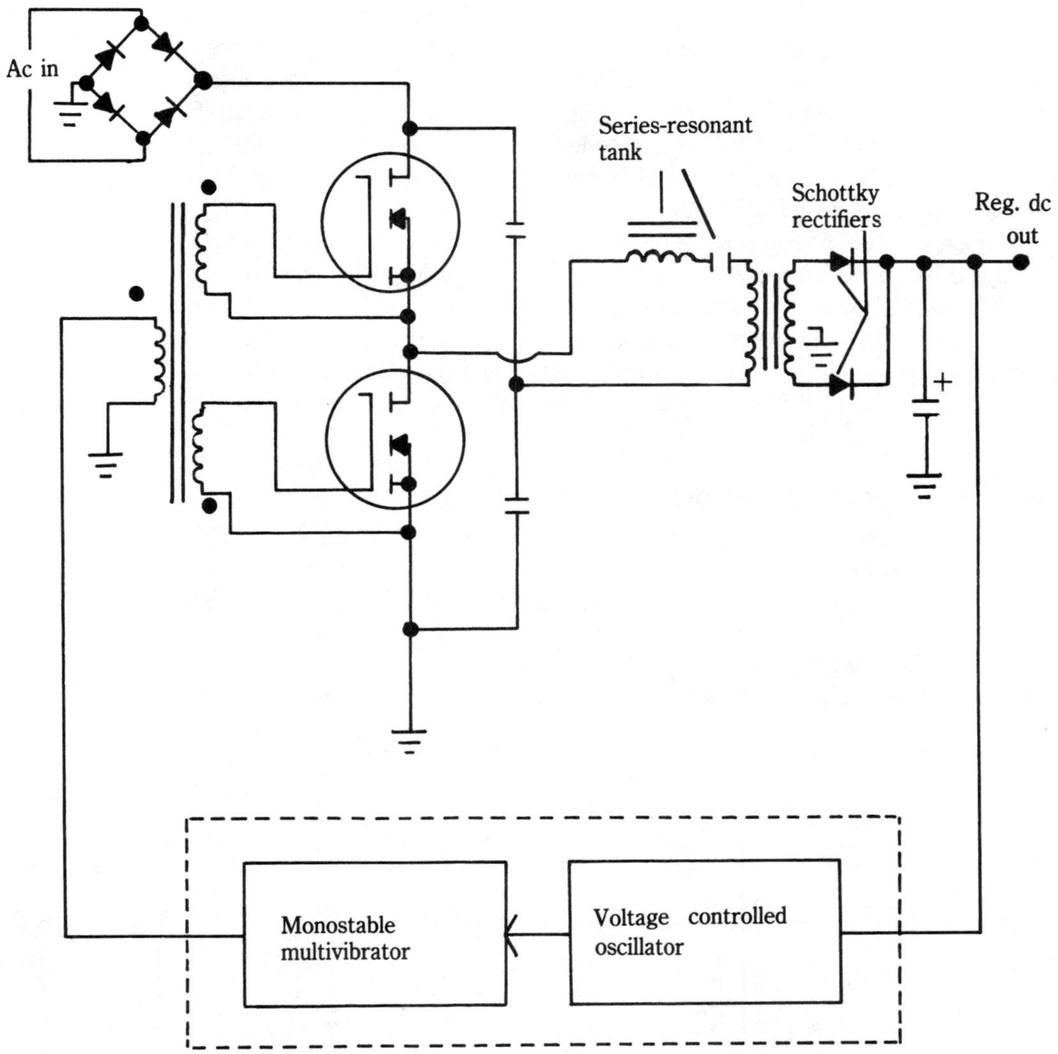

1-14 Basic scheme for one form of resonant-mode power supply.

control circuitry. Also, high-frequency techniques must be used throughout. For example, the output rectifiers should not pose reverse recovery losses, Schottky diodes usually fill this requirement. Other rectifiers, such as MOSFET synchronous rectifiers, or special pn diodes also can be satisfactory. Various adaptations of this half-bridge circuit can be advantageously used throughout the several hundred kHz to several MHz range. The parasitic diodes of the power MOSFETs are not shown in the diagram so attention can be focused on the key attributes of the circuit.

Although it is possible to accomplish regulation by centering operation on either the low-frequency or the high-frequency slope of the resonance curve, it generally would be more convenient to use the low-frequency slope. The best compromise in a number of operating conditions results when the Q of the resonant circuit is just high enough to apply a good approach to a sine wave across the primary of the output transformer under full-load operation. Inadequate Q tends to subject the power MOSFETs to high peak currents. Excessively high Q makes it difficult to extract needed power from the capacitor. Low Q is obtained in series-resonant circuits by tending toward low inductance and high capacitance. In the design and operation of this type of regulated power supply, it should be ascertained that neither the power MOSFET drive transformer, nor the output transformer saturate.

Three pioneers in the marketing of control ICs for resonant-mode power supplies have been Cherry Semiconductor Corp., Unitrode Integrated Circuits, and Gennum Corp. of Ontario, Canada. The major American semiconductor firms are bound to follow suit.

Resonant-mode operation in a forward converter

The simplified circuit shown in Fig. 1-15 is that of a forward converter. This much used switcher circuit is readily changed into a resonant-mode circuit via two modifications. One modification is the addition of a small capacitor (shown in dashed lines in Fig. 1-15). The other modification is application of a constant duration rectangular wave to the base of the switching transistor of such a repetition rate that the leakage inductance of transformer T1 is resonated by the added capacitor, C.

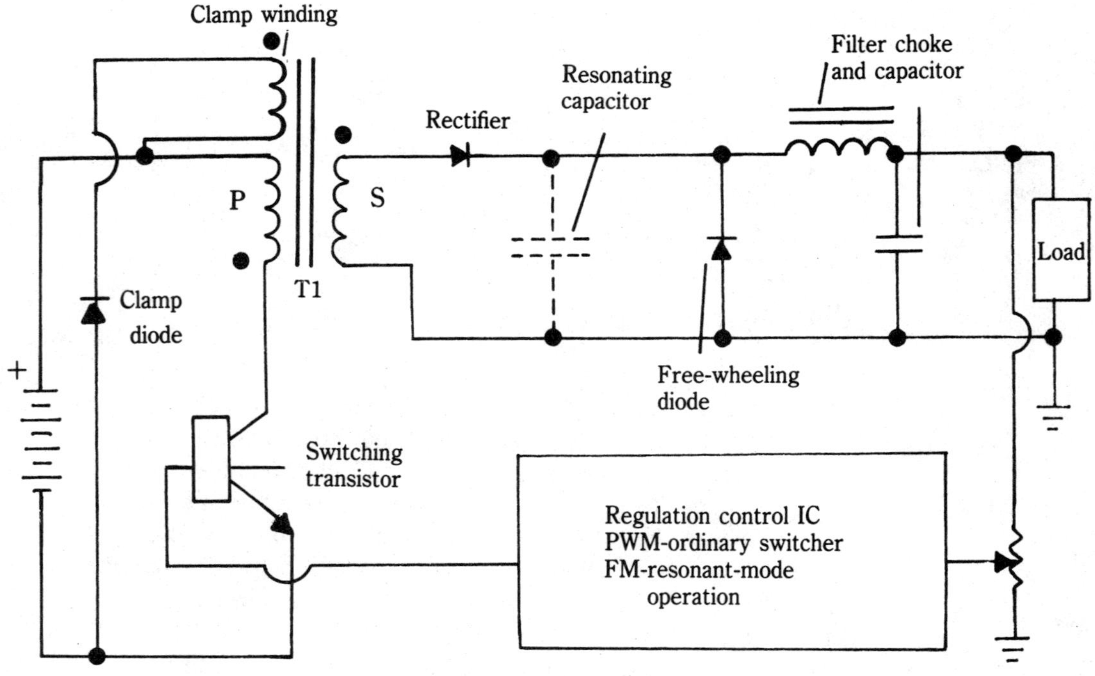

1-15 Obtaining resonant-mode operation from a forward converter.

When the forward converter is operated as a conventional circuit, a PWM wave is applied to the base of the switching transistor and the frequency is much lower than when the described modifications produce resonant-mode operation. It goes without saying that the PWM control IC (not shown in the simplified schematic) must be replaced by a dedicated FM-type operation when resonant-mode operation is desired. You also must assume that the switching transistor will continue to work well at the higher frequency. This probably will be the case, because the transistor will now carry half sinusoids of current, rather than a near square wave (or near rectangular waves). This is, in itself, a more relaxed burden for the transistor, and will be manifested as a worthwhile reduction in switching losses.

It can be appreciated that there is more than meets the eye to such a resonant mode converter. The resonating capacitor, C, is not used to tune the secondary of transformer T1, as a casual inspection might suggest. Rather, the "invisible" leakage inductance is series resonated by capacitor C. Designing transformer T1 to satisfy the usual parameter requirements, and to additionally exhibit a certain leakage inductance, might not be an easy task. A semiempirical method where an already satisfactory transformer (for the conventional forward conversion mode) is series resonated at some high frequency by experimenting with capacitor C can be a practical way of attaining resonant-mode operation.

When operating in the resonant mode, regulation is obtained by varying the frequency of the pulses applied to the base of the switching transistor. This is because the circuit is working on the slope (usually the low frequency slope) of the resonance curve established by T1 leakage inductance and capacitor C. A little contemplation will show that stray circuit parameters that ordinarily prove troublesome in the PWM mode can be used to advantage in the resonant mode. Also, in this particular application of resonant-mode operation, a half sinusoid, rather than sustained ringing, is produced in the resonant circuitry.

The FM control technique is also known as ***pulse-density modulation***, or PDM. This distinguishes it from PWM which is commonly used in lower frequency switching supplies.

The current-driven sine wave inverter

Not all sine wave power supplies are known as resonant-mode types. Although these circuits tend to be "kissing cousins," there are design and operational differences despite aspects they share in common. Sometimes, however, classification by name seems to be rather arbitrary.

A case in point is the current-driven sine wave inverter. Although the name of this circuit emphasizes operating mode rather than resonance or shock excitation, these aspects of circuit behavior are nonetheless present. It must be stated, however, the constant current inverter was not developed with an eye to high frequency operation. There is no obstacle to its adaptation to high frequency operation and it is likely that this circuit will, indeed, receive consideration for higher switching rates. (Do not be surprised if the curcuit falls under classification as a resonant-mode switcher.)

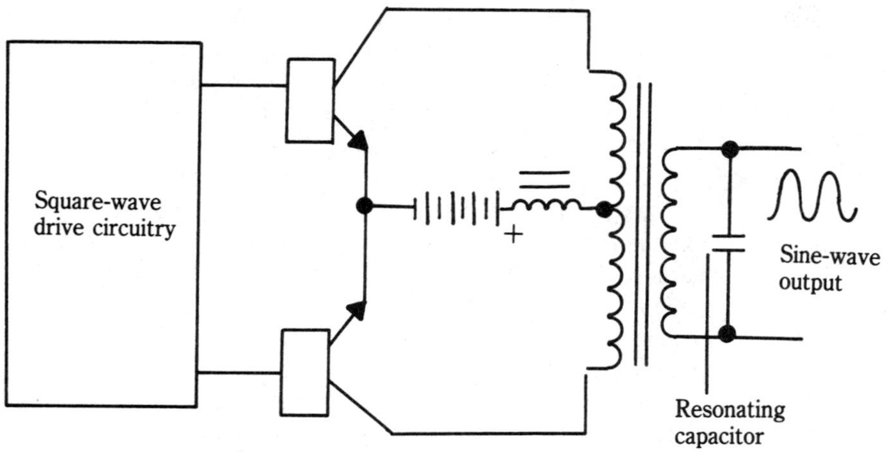

1-16 Basic setup for a current-driven sine-wave inverter.

The basic circuit of the current-driven sine wave inverter is shown in Fig. 1-16. Its salient circuit features are the insertion of an inductor in one of the DC leads, and the use of a tuned output circuit. These modifications from conventional inverter circuits, result in quite different performance; some of these differences can be very beneficial.

The conventional (voltage-driven) inverter has limited capacity for handling inductive loads, and even less for capacitive loads. The current-driven sine wave inverter can handle a wide range of reactive loads. The conventional inverter is notorious for its radiated and conducted EMI, whereas the current-driven sine wave inverter is nearly free of such generated interference. Although conventional inverters have long been used to drive motors, the current-driven inverter generally is superior for this task because of its constant current characteristic; there is no excessive inrush current at standstill, even though full starting torque is available. Also, temperature rise from hysteresis is less because of the absence of harmonics in the applied sine wave. These factors not only improve motor performance, but extend motor life.

Power MOSFET high-efficiency half-wave rectification

The rectifier circuit of Fig. 1-17 utilizes a power MOSFET and a voltage comparator, or operational amplifier, to achieve lower forward voltage drop and overall higher rectification efficiency than is provided by diode rectifiers. Additionally, it will operate at higher voltages than those realized from Schottky diodes. At the same time, its frequency capability exceeds that of silicon rectifying diodes. A good approach to the operating principle of this circuit is to first state what it is not.

Notwithstanding rather suggestive circuit topography, this rectifier circuit is not a so-called *precision rectifier*. A precision rectifier uses an op amp to nearly

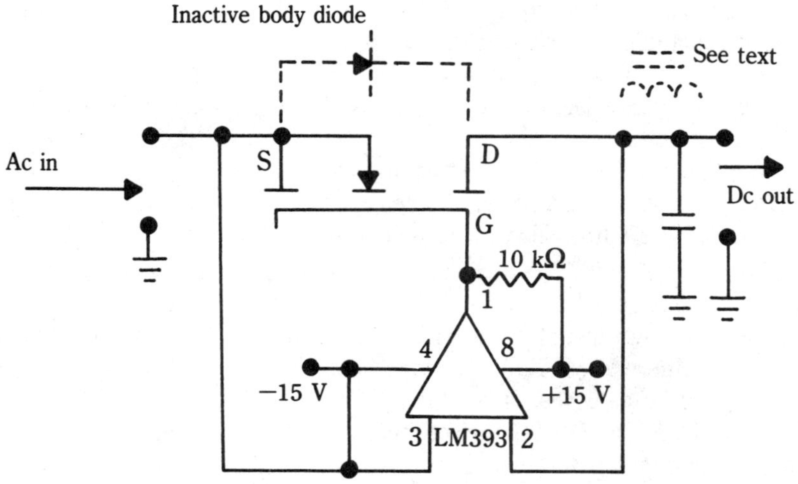

1-17 Electronic rectifier circuit using low R_D power MOSFET.

neutralize the voltage gap of rectifier elements. Thus, the 0.7 volt gap of a silicon diode can be made to appear as 0.007 volt with the aid of the 100:1 voltage gain of an op amp. The drawback of precision rectifiers is that they are essentially small signal circuits, unable to handle currents needed in power applications.

Neither is the circuit of Fig. 1-17 an ordinary synchronous rectifier, notwithstanding the control voltage applied to the gate of the power MOSFET. Although the current through the power MOSFET is passed and blocked in synchronism with the inputted AC wave, the timing is the brainwork of the voltage comparator. It seems best to call this circuit an *electronic rectifier*.

When in proper operation, rectification is not performed by the internal body diode of the MOSFET. Rather, forward conduction current is carried by the source drain circuit of the MOSFET, and the source-drain voltage drop developed is lower than the 0.7-volt threshold of the body diode. Therefore, the body diode is electrically inactive.

Finally, the alluded forward conduction of the n-channel MOSFET occurs with unconventional polarity, the source being positive with respect to the drain. (In a sense, the MOSFET is the device capable of "bipolar" operation, rather than bipolar transistors.)

In light of these facts, a straightforward description of circuit operation now can be given. The voltage comparator is so connected that it delivers conductive bias (positive voltage) to the gate of the MOSFET whenever it senses that source voltage is positive with respect to drain voltage. Otherwise, the conductive state of the voltage comparator deprives the MOSFET's gate of the needed ON signal. Thus, no current is passed through the MOSFET when its drain is properly more positive than its source. The overall process constitutes very efficient rectification.

The output filter capacitor in this circuit can, in some cases, be a crucial component. This is because of the peaked current demand of such a primitive filter; the peaked current is accompanied by peaked voltage drop across the MOSFET, and this, in turn, causes conduction in the body diode. Higher conductive and switching losses will be introduced into the rectification process. This situation is more likely to be encountered with sine waves than with square waves. In any event, a possible remedy is to empirically determine the smallest capacitor that can be used. A better approach, especially if good filtering is needed, is to insert a choke ahead of the capacitor. Such an L-section filter protects the MOSFET from high capacitor currents.

The power MOSFET should be selected so that the product of load current and R_D is comfortably less than 0.4 volts, allowing for the fact that R_D will be higher at operating than at room temperature. An excellent type for this purpose is the Motorola MTH40N05, which exhibits an R_D of about 0.028 ohm while passing 10 amperes. Also, International Rectifier Corp. and Samsung offer the IRF250 and IRF140, which have an R_D of 0.085 ohm at higher than usable current ratings for this rectifier circuit. These two power MOSFETs would work well supplying a four ampere load. Also available from these two vendors is the IRF150 with R_D = 0.055 ohm, and usable in this rectifier circuit for a six ampere load current.

Synchronous rectification with complementary power MOSFETS

An interesting and useful application of power MOSFETs is shown in Fig. 1-18. This is the circuit of a full-wave synchronous rectifier. It makes use of a bridge configuration like that commonly employed with diodes. Current is passed by each MOSFET when its gate receives appropriately polarized voltage. Note that two of the MOSFETs are n-channel, and two of them are p-channel

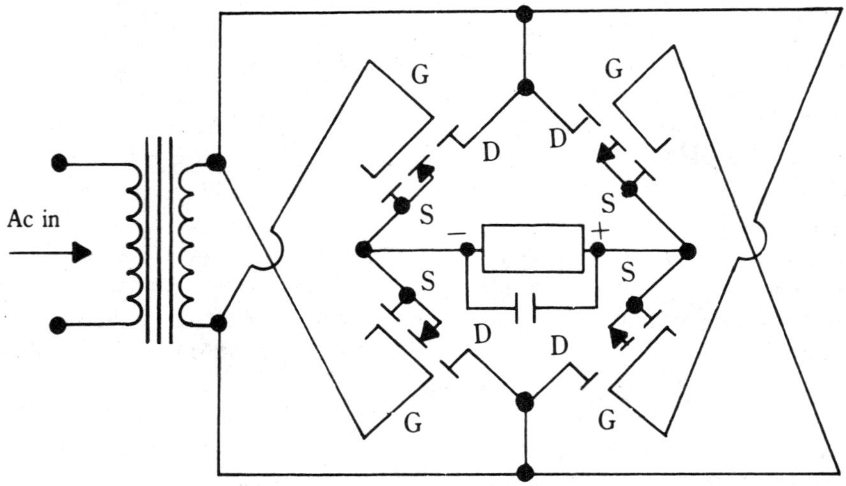

1-18 Synchronous rectifier using n-channel and p-channel MOSFETs.

devices. Not shown in the diagram are the internal body diodes of the MOSFETs because in proper operation of this rectifier bridge, the body diodes are electrically inactive. To draw them in would add confusion to a circuit that already is a brain teaser.

It might seem that the brief discussion here suffices to depict the manner in which this circuit works. It would seem that active elements have been substituted in place of conventional rectifying diodes, and that these active devices conduct at appropriate times to produce rectification. Indeed, this description is true. An important point has been overlooked thus far: It turns out that if you deal with the normally encountered drain source voltage polarity of the four MOSFETs, each one would be seen to be biased off. Apparently making matters even worse, it is seen that when turn on gate bias exists, the polarity of the drain source voltage is reversed from what it is in normal circuits. All told, it would appear that no rectification can take place because individual MOSFETs never experience ''correct'' polarization of their electrodes.

In an n-channel MOSFET, it is true that the drain is commonly positive with respect to the source. However, the device would be equally responsive for negative drain and positive-source operation except for one thing. That one thing is the internal body diode, for it would go into forward conduction if the drain were made negative with respect to the source. The same reasoning applies to the p-channel MOSFET in which the drain is operated positive with respect to the source.

In the circuit shown in Fig. 1-18 the conduction of the body diodes is prevented by deliberately using MOSFETs with very low values of R_D. The basic idea is to maintain the drain source voltage drop less than about 0.7 volt so that the body diode in the MOSFETs cannot conduct. This is why it was stated that the body diodes are inactive. It is obvious that the MOSFETs must be selected carefully against the desired load current to achieve synchronous rectification.

A natural question arises as to what would happen if the voltage drop across the MOSFETs exceeded a nominal 0.7 volt. Although it has been pointed out that synchronous rectification would no longer occur, it should be appreciated that rectification would still occur, because the body diodes would then become active and behave as ordinary rectifying diodes. Actually, for the body diodes to take-over rectifying operation, the peak voltage of the drop across the drain source electrodes would have to be considerably more than 0.7 volt—perhaps several volts or more. The point here is that the transition from synchronous to nonsynchronous (conventional) rectification is smooth. The main evidence of such transition would be increased voltage drop in the rectifier circuit.

This brings you back to the reason for using synchronous rectification in the first place. Such a rectification system, employing low R_D MOSFETs, is able to compete with, or exceed, the rectification efficiency of germanium and Schottky diodes. At the same time, the MOSFETs are good at high frequencies, and can perform well at higher voltages than Schottky diodes. Once you grasp the principle of operation, other rectifying circuits, such as full-wave centertap, or half-wave arrangements can be readily devised.

When the synchronous rectifier is used with square waves, it is easy enough to arrange the circuit so that the gates of the MOSFETs will receive a sustained 10 or 12 volts throughout a half cycle. For operation with sine waves, Zener diodes can be placed across the gate source terminals and sufficient AC voltage can be applied to the gate circuits so that the rectified waveform is not appreciably narrowed. Such Zener diodes should generally provide 12 to 15 volts to the gate source circuits to ensure hard saturation of the power MOSFETs.

Besides a low forward voltage drop, another feature of this rectification scheme is that negligible losses result from reverse recovery characteristics or from reverse leakage. This is because the ordinary behavior of MOSFETs prevails—no minority carrier storage effects, nearly complete cutoff at below threshold gate bias, and excellent high frequency operation.

A choke input filter might be necessary to keep capacitor current peaks from causing conduction of the body diodes. The situation is less crucial for this full-wave circuit, however.

Full-wave synchronous rectifier for high-frequency square waves

The full-wave synchronous rectifier shown in Fig. 1-19 can prove more effective for the rectification of high-frequency square waves than Schottky diodes, which often lose efficiency from reverse leakage. This is particularly true at high temperatures, and when handling more than 10 or 15 volts. It is important, however, that appropriate power MOSFETs are used. Semiconductor firms

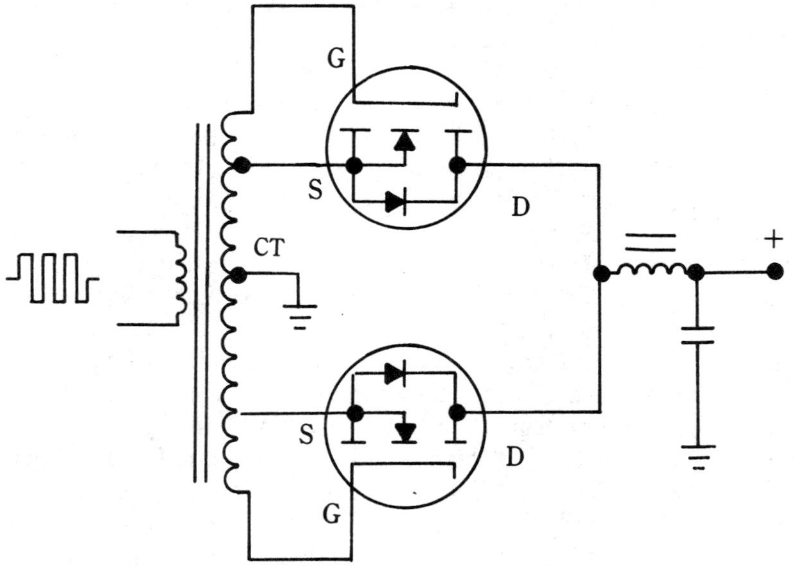

1-19 Full-wave synchronous rectifier.

now make synchronous rectifier optimized devices that have inordinately low R_D values. It is a good idea to look for these in the vendor's catalogs.

As with other synchronous rectifier circuits using power MOSFETs, the circuit shown in Fig. 1-19 operates these devices with polarization opposite to that used in most MOSFET applications. This is permissible because the power MOSFET, unlike the bipolar transistor, has bilateral conduction characteristics. It is important that R_D be very low, otherwise the intrinsic diode will go into forward conduction, thereby rendering the circuit inoperative as a synchronous rectifier. Synchronous rectification is brought about because the gates receive positive turn-on voltage for only the half cycle while the drain source circuit is reverse biased. When the drain source circuit is forward, or conventionally biased, the gates are reverse biased and no conduction takes place. (At this time, the intrinsic diode is also reverse biased, so there is no danger of it getting into the act; indeed, with proper operation, the intrinsic diode is always inactive.)

The taps (other than the centertap) on the transformer generally should be arranged to provide 12 or 15 volts between source and gate to ensure hard turn on during conduction time.

This circuit merits consideration for use in the several-kHz to several-MHz range, or where the leakage problem with Schottky diodes is troublesome. There is nothing to prevent the use of this scheme at lower frequencies except that it might not prove cost-effective in comparison with conventional rectifier diodes. Another applications area is for two and three-volt power supplies. For such service, you must acquire the best-suited power MOSFETs, and a probe of the factory's specialty items is likely to prove rewarding.

Although the circuit is at its best with square waves, very good performance attends the use of PWM waveforms as well. Sine wave operation would require some experimentation with the gate signal to maintain drain source conduction over a half cycle.

A practical example of a resonant-mode power supply

After your investigation of high-frequency techniques, it is fitting to consider a practical design. A typical resonant-mode power supply is shown in Figs. 1-20 and 1-21. The heart of this, and similar circuits that can be scaled up and down in power level, is the Gennum GP605 resonant-mode control IC. The Gennum Corporation has been a pioneer in the development of resonant-mode power supplies operating in the several-hundred kHz to several-MHz region, where power-to-volume ratios in the vicinity of 25 watts per cubic inch can be attained.

As with PWM regulation techniques, it is much better to work with a dedicated IC than attempt to implement the control function via discrete devices. This is especially true at the higher switching rates allowed by the resonant mode technique. A nice thing about the dedicated control IC is that it comes equipped with all manner of bells and whistles; these often prove stumbling

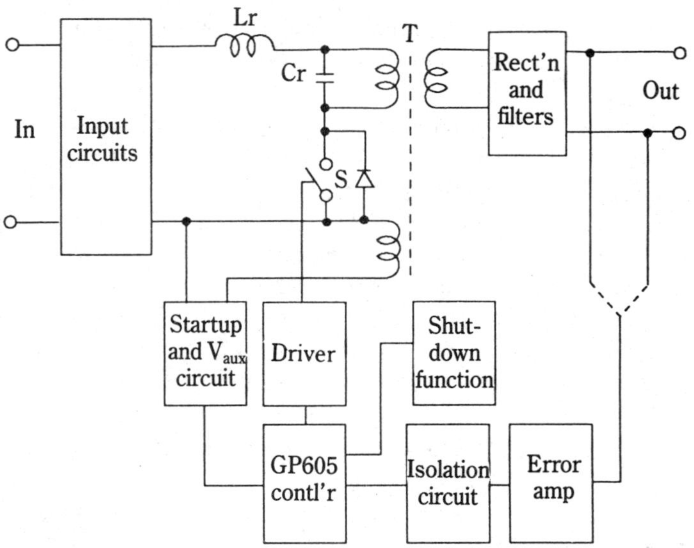

1-20 Block diagram of the resonant-mode power supply.

blocks for the designer, who often surmises that various auxiliary functions can be "thrown in" as a last-minute thought.

The block diagram of the GP605 IC is shown in Fig. 1-22. Note that the "bare bones" portion of this control circuit is the *voltage-controlled oscillator* (VCO) and the monostable circuit. Through the action of these two function blocks, the resonant circuit in the power supply is impressed with a constant duration pulse train that varies in repetition rate in accordance with a sensed sample of the output voltage of the power supply. The impedance curve of the resonant circuit enables regulation to be achieved because the average ON time of the power switch can be varied to maintain constant voltage at the output of the supply.

In addition to this basic operational format, there are a number of options that serve various conveniences and protective functions. If pin 10 (SEO) is grounded, complementary output is obtained for push-pull applications. If pin 10 is open, single-ended output is delivered at double the frequency. Under this condition, OUT A and OUT B can be connected together for higher drive capability.

Remote shutdown is under control of pin 1 (RSD). Logic low on pin 1 shuts down the GP605. Releasing the pin causes the GP605 to go into soft start.

Soft start operation is provided by connecting an appropriate capacitor from pin 12 (SS) to ground. This user option is particularly useful because you should use slower startup at higher power levels. Whereas 50 milliseconds of such graduated delay might be suitable for the 100-watt level, a 500-watt supply would be safer to incorporate about 500 milliseconds. Interestingly, soft start in the resonant-mode supply operates by starting the switching rate at a very low frequency and gradually increasing it to the operating rate. This is a smooth transient-free method of placing the power supply in operation.

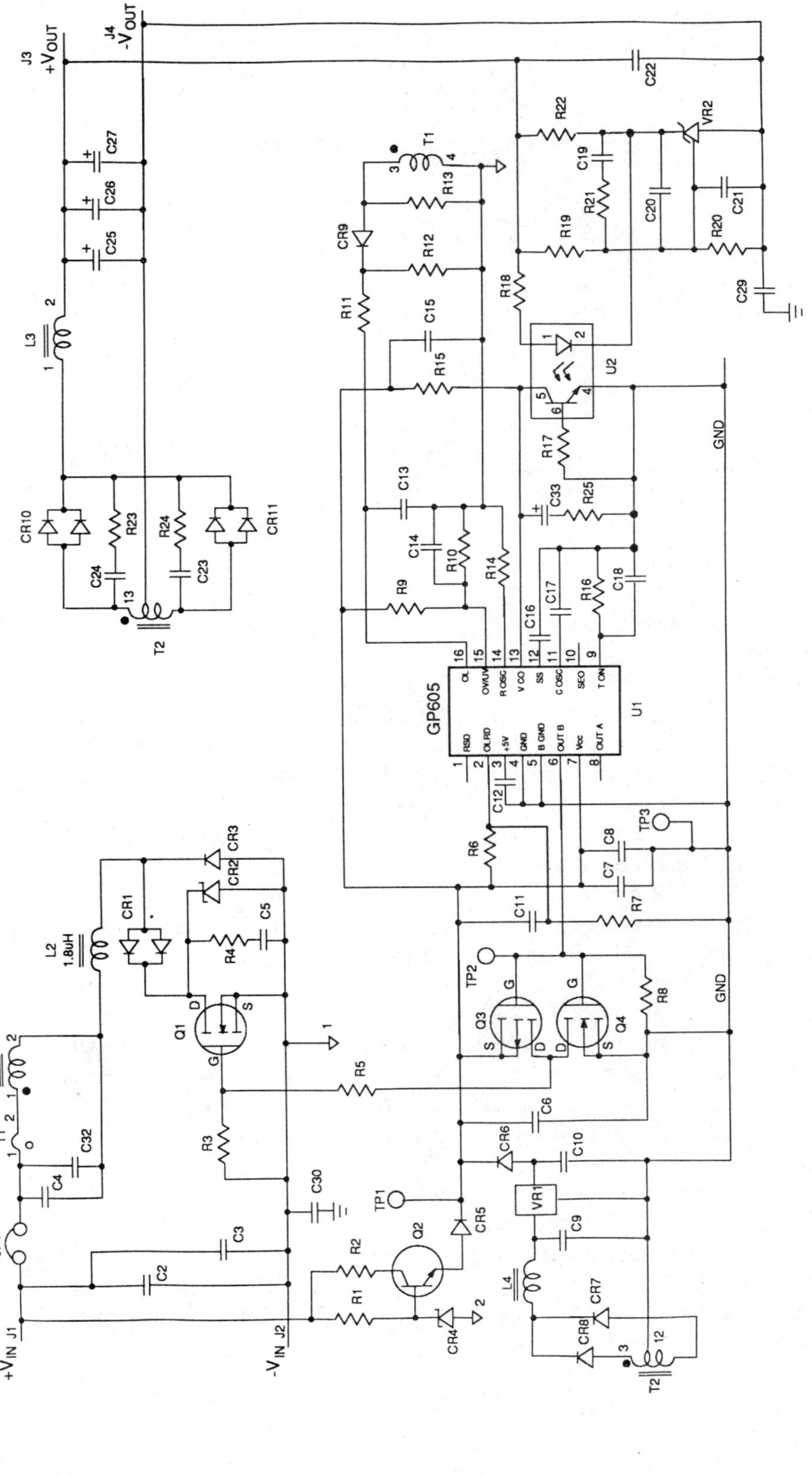

1-21 Schematic diagram of 48V/5V, 20 A, resonant-mode supply.

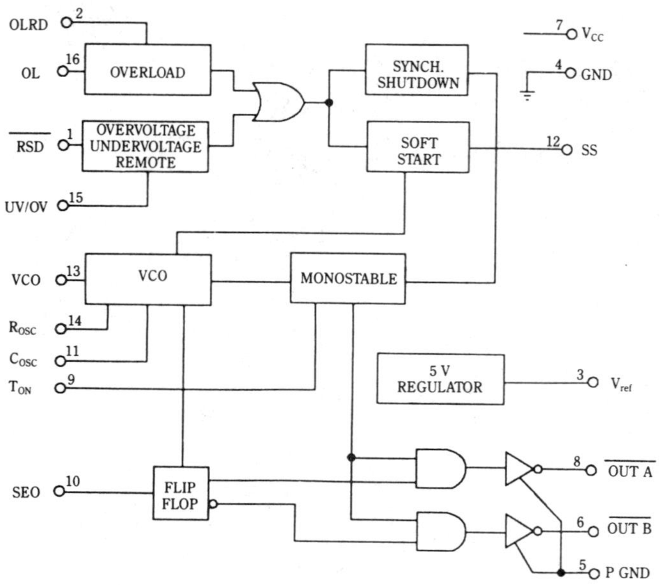

1-22 Block diagram of the GP605 resonant-mode controller.

Yet another protective feature is the overvoltage – undervoltage shutdown provided by an appropriate external network connected to pin 15 UV/OV. This is accomplished by a window comparator within the chip.

With regard to the various shutdowns that might intentionally or unintentionally occur, subsequent startup is always under control of the soft start function. Moreover, shutdown never takes place in the middle of an ON pulse, which could generate destructive transients. It can be seen easily that the availability of a control IC, such as the GP605, with "designed in" features goes a long way in making the implementation of basic concepts such as resonant-mode operation at high switching rates practical.

The maximum frequency of the VCO is governed by an external capacitor connected to pin 11, (C_{OSC}), and an external resistor connected to pin 14 (R_{OSC}). Also, the fixed duration of the monostable ON pulse is set by an external RC network connected to pin 9 (T_{ON}).

The time constant of an external RC network connected to pin 2 (OLRD) determines how long the supply remains shutdown from an overload. The result of this temporary shutdown is that the power supply protects itself in a "hiccup" mode; it repetitively seeks normal operation through the soft start cycle. Normal operation automatically establishes itself once the cause of the overload is removed. Sensing of the overload is done by a voltage applied to pin 16 (OL).

Shifting attention back to the overall schematic diagram of Fig. 1-21, it will be seen that all power supply functional blocks fit together in a straightforward fashion. Note in particular, the resonant tank (L2, C4, C32), the driver (Q3 and Q4), and the optoisolator (U2) in the feedback loop. These components optimize performance at the high switching rate, in the vicinity of 600 kHz.

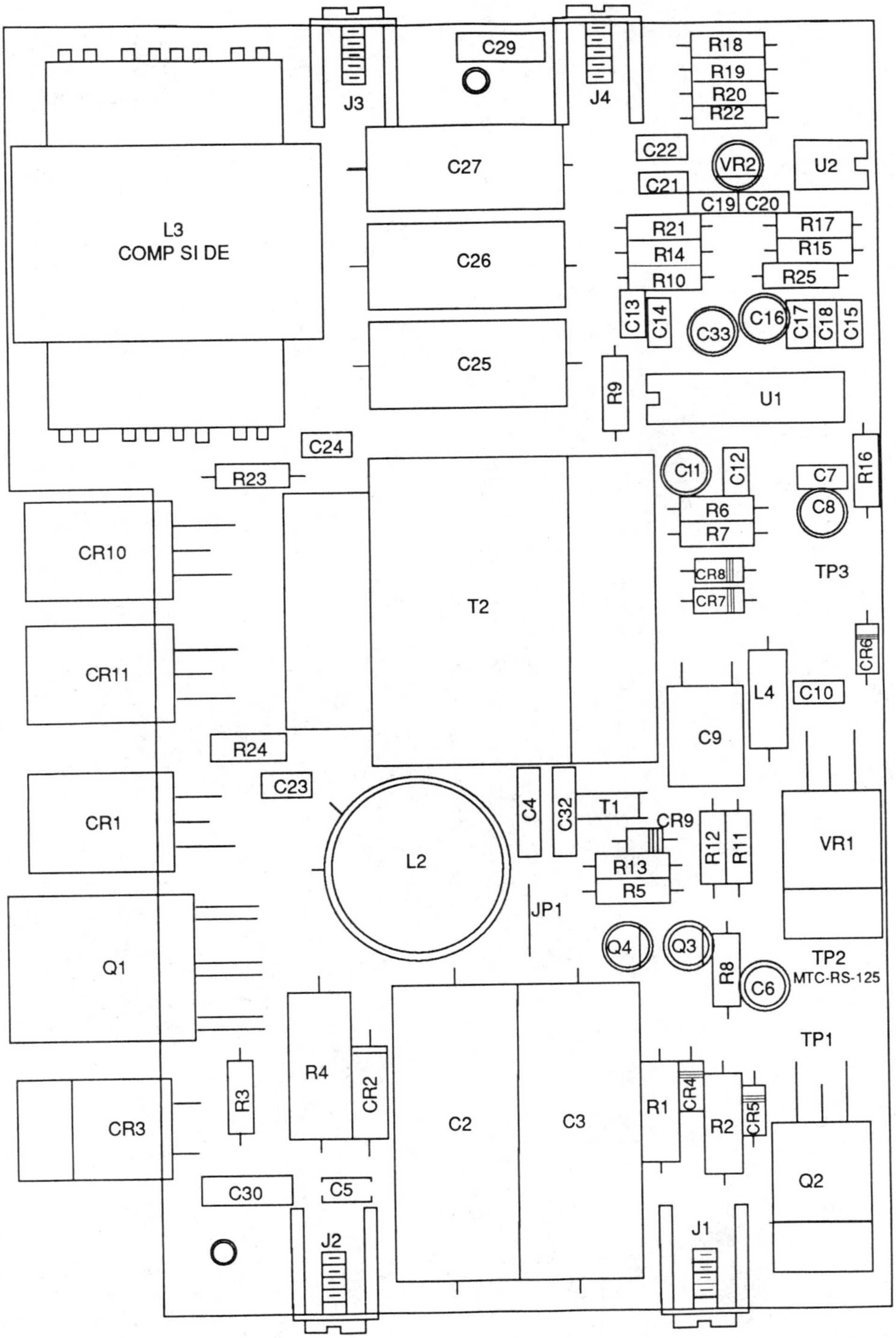

1-23 Printed-circuit board layout of resonant-mode power supply.

Table 1-1. Bill of materials for resonant-mode power supply

Part no.	Description	Part no.	Description
C2, C3	6.0 μF 100 V ±10% Electrocube 230BIB605k	R5	2.2 Ω 1/4 W 5%
C4, C32	0.022 μF 200 V NPO ±5% KEMET C340C223J2G5CA	R6, R7	300 kΩ 1/4 W 5%
C5	680 pF 200 V NPO (COG) ±5% KEMET C323C681J2G5CA	R9	3.6 kΩ 1/4 W 5%
C6, C8	22 μF 16 V tant. dipped cap. Sprague 199D226X9016DA1	R10	1 kΩ 1/4 W 5%
C7, C10, C15, C19, C20, C22	0.22 μF 50 V Z5U ±20% Sprague 1C10Z5U224M050B	R11	220 Ω 1/4 W 5%
		R12	36 Ω 1/4 W 5%
C9	47 μF 20 V tant. dipped cap. Sprague 199D476X9020EE2	R13	22 kΩ 1/4 W 5%
C11, C16, C33	4.7 μF 16 V tant. dipped cap. Sprague 199D475X9016BA1	R14	26.1 kΩ 1% 1/4 W
C14, C21, C12	0.001 μF 100 V ±10% Sprague 1C10X5R102K100B	R15	3.9 kΩ 5% 1/4 W
C13	0.022 μF 50 V ±20% Sprague 1C10Z5U223M050B	R16	10 kΩ 1% 1/4 W
C17, C18	100 pF 100 V ±5% Sprague 1C10C0G101J050B	R17	1 MΩ 5% 1/4 W
C23, C24	0.0022 μF 100 V COG Sprague 1C10C0G222T050B	R18, R22	470 Ω 5% 1/4 W
C25, C26, C27	220 μF 10 V 20% solid tantalum KEMET T262D227M010MS or CSR21C227KM	R19, R20	1 kΩ 1% 1/4 W
		R21	10 kΩ5% 1/4 W
C29, C30	0.0022 μF 500 V cer. disk Sprague 5TSD22 or 5GAD22	R23, R24	51 Ω 1 W 5% RCD RSF1A
CR1	Schottky diode TO-220AB DUAL Amperex BYV4335 or Motorola MBR2035CT	R25	100 Ω 5% 1/4 W
CR2	Zener diode 170 V ±10% 5 W IN5385A	L2	2 μH ±5% Core: T68-2D Micrometals 10 turns #30 AWG × 5 (wire length = 10 inches)
CR3	UFRD 15 A 200-V-to-220 AC Single Amperex BYV29-200 or Motorola MUR1520	L3	MTI-125-02-02 GAP = .012 Multisource Technology Inductor

CR4	Zener diode 12 V ±5% 1/4 W IN4699	L4	330 μH 0.6 ohm Inductor AL0410-331K Northeastern Electronics (315) 455-7561
CR5, CR6	IN4001 IA 50V	T1	Current sense transformer: Primary: 1 turn #18 AWG Secondary: 50 turns #34 AWG Core: Ferroxcube 1041CT060/3E2A
CR7, CR8	UFRD AXIAL 1A 200 V Amperex BYV27-200 or Motorola MUR120		
CR9	IN4148 Signal diode	T2	MTT-125-DC-06-02-06C Multisource Technology transformer
CR10, CR11	Schottky diode T0-220, DUAL Motorola MBR2545CT or Amperex BYV 43-45	VR1	LM7812 12 V regulator
Q1	MOSFET IRFP250 International Rectifier T0-247AC	U1	GP605 Gennum resonant mode controller
Q2	TIP 29C TI bipolar transistor	U2	CNY17-4 TRW opto-coupler
Q3	MOSFET VP0104N3 Supertex or Plessy ZVP2106A	VR2	TL431 CLP TI shunt regulator
Q4	MOSFET VN1306N3 Supertex or Plessy ZVN3306A	Mounting hardware	4 -40 nuts and flathead bolts, T0 - 220 insulating bushings, Bergquist insulators, K4 - 90, K4 - 35, Washers for PCB spacing, Baseplate, Heatsink, PCB, and Keystone #8190 terminals with screws
R1	180 Ω 1/2 W 5%		
R2	3.9 kΩ 1/2 W 5%		
R3, R8	51 kΩ 1/4 W 5%		
R4	47 Ω 3 W 5% RCD RSF2B or Clarostat VC-3D		

Gennum Corporation assumes no responsibility for the use of any circuits described herein and makes no representations that they are free from patent infringement.

Figure 1-23 illustrates the layout of the supply. This is not shown as a "how-to-do-it" instruction; rather, it is intended to provide a general guidance in the location of components in terms of good high-frequency practice. You can perceive that power supplies are beginning to resemble rf circuits used in receivers and transmitters. Indeed, the same regard has to be given to such matters as radiation, stray capacitance, and parasitic inductance. That is why a well thought out layout together with fabrication via a PC board are almost requisites of successful operation.

Table 1-1 depicts the bill of material, or parts list. A cursory study of this list reveals no need for inordinately tight tolerances. This should dispel the often encountered notion that circuits involving resonance must be crucial to place in operation and/or be difficult to reproduce.

Enhanced efficiency from the linear voltage regulator

A hybrid regulating system made up of a switch-mode supply followed by a linear post regulator combines some of the salient features of both types of regulation. For all practical purposes, the post regulator is simply added to what otherwise would be the output of the switching supply. This is particularly easy to do because the linear regulator is outside of the regulating feedback-loop of the switching supply. Such a scheme provides a smoother output than is attainable from the switching supply alone. At the same time, the overall system is much more efficient than could be expected from a linear regulator alone. This is primarily because the linear regulator is not exposed to high input voltages—that is, it has the benefit of a preregulator (the switching supply).

There is another approach to the objective of getting a good compromise between efficiency and close regulation or quiet performance (in the electrical sense). In this alternative scheme, the roles of the two regulators are reversed; the linear regulator becomes the main functional block, with the switching circuit playing an auxiliary, but important role. Thus, in Fig. 1-24, the LT350A linear voltage regulator functions as it would if it were the only active device, but with one exception—the voltage drop across its input-output terminals is maintained at V_Z, a low voltage. This is accomplished by the LT1018 and the 2N6667, and associated circuitry. These circuitry elements constitute a switch-mode supply that varies its duty cycle in such a way as to regulate V_Z. Although the LT350A develops dissipation in the manner of all linear regulators, such losses are minimal because V_Z is not much greater than the minimum value needed for operation.

The actual value of V_Z is 3.7 volts. Thus, the full-load dissipation in the linear regulator is nearly equal to the product of full-load current and V_Z, or $3 \times 3.7 = 11.1$ watts, which cannot be exceeded regardless of the input or output voltage of the system. The self-excited oscillation of the circuit is in the vicinity of 5 kHz when the output is 15 volts and 3 amperes, so switching losses tend to be low. The hysteresis of the LT1018 comparator is set at about 80 millivolts p-p by the 68-pF, 1-megohm combination. Startup of oscillation is ensured by the 2.2K resistor.

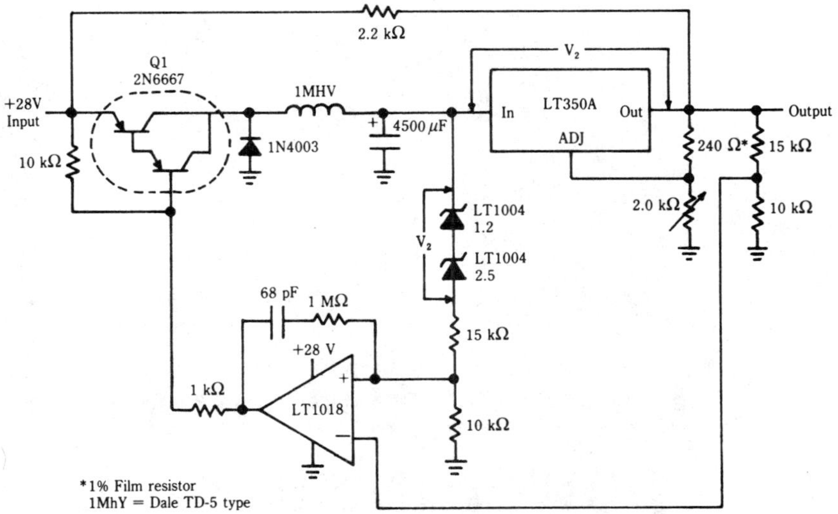

1-24 Unique use of switching and linear regulating techniques.

The basic concept of maintaining a low voltage drop across a linear regulator merits consideration where the high efficiency of the switching supply is not needed, but where the inordinately low efficiency of the linear regulator alone cannot be tolerated.

2

A look at the newer power devices

POWER SUPPLY EVOLUTION HAS LONG BEEN CENTERED AROUND BIPOLAR power transistors, power MOSFETs, and thyristors. Included in the bipolar stable has been the power darlington. Thyristors, besides involving the SCR, have also included the triac and the gate turn off (GTO) thyristor. These devices have been contending for superiority in switching speed, current, voltage, and power capability, ease of implementation, reliability, and of course, cost. Sometimes immediate benefits of an improved design were not readily forthcoming—it was necessary to wait until component manufacturers offered products that could advantageously work at higher frequency, voltage, or current. At other times, the advanced components made their debut first, challenging the semiconductor companies to catch up.

You can peruse through yesteryear's technical literature and find many instances in which conventional wisdom pointed to the rapid demise of one or another power device. It certainly appeared that the power MOSFET had sealed the doom of the bipolar power transistor. What actually happened, however, was that the bipolar transistor was unwilling to roll over and die. Rather, it quite successfully embarked upon a program of self-improvement; it upgraded its performance parameters to the extent that designers were forced to give serious consideration to the choice between MOSFETs and bipolars. For example, the bipolar transistor soon enough exceeded its perceived top switching rate of 25 kHz, or so, enabling efficient and cost-effective regulators to operate in the vicinity of 100 kHz.

At the same time, power MOSFETs have been busy overcoming their chief drawback—high internal resistance at high voltages. Indeed, this parameter (R_D) has so dramatically decreased since power MOSFETs first appeared on the market that designers who haven't kept up with the times do themselves an injustice by rejecting high voltage MOSFETs on the basis of their outmoded model R_D.

Nor has the power darlington slumbered in the face of the fray. Those annoyed by the drive power needed by single element bipolar transistors find that there is an appreciably wide application area where the power darlington, with its low drive requirements, becomes the most appropriate power device. The darlington, because of its monolythic fabrication, can be considered a power IC. This is interesting because power ICs in the form of "smart" power modules are increasingly asserting themselves as substutites for the simpler devices, often providing useful control and protection functions. Originally a "power op amp," these smart power modules now offer to lighten the designer's task in guarding both supply and load against abusive circuit conditions.

During the ongoing contest among the various types of transistors and the emerging power modules, designers felt they could leapfrog the performance of these devices, especially at high power levels, if only a way could be found to turn off an SCR via its gate. From this hope was born the GTO thyristor, which could have its conduction terminated by a negative gate pulse. Popularity waxed and waned for this device several times within a like number of years; finally, pursuing its further development was left to overseas technology. The Achilles' heel of the device has been its need for inordinately large turn-off current, usually about a third of its rated full-load current.

It has finally been realized that one type of device doesn't appear destined to suit all requirements of speed, voltage, current, power, efficiency, or cost. Rather, each device displays optimum behavior in certain realms. At the same time, there is much overlap in which competent circuit design can extract desirable performance from more than one device.

Moreover, all the while improvements have been wrought in the traditional power devices (including diodes and rectifiers), new devices have made their debut. Some of these have intriguing features and will be investigated now.

Improved performance from nearly-mature devices

Although you are primarily interested in investigating new types of semiconductors, the improvements in the more familiar ones have, in some instances, been so dramatic as to merit consideration. For example, the active drain source resistance, R_D, of power MOSFETs has, because of its relatively high value compared to the equivalent parameter of bipolar transistors, often robbed the MOSFET of its otherwise competitive edge for low- and medium-frequency operation.

However, a very appreciable reduction in R_D has been accomplished over early models; those who continue to think in terms of the initially available devices do themselves an injustice. Figure 2-1 depicts the "learning experience" of 50 to 100 V power MOSFETs over the 1977 to 1988 period. R_D is

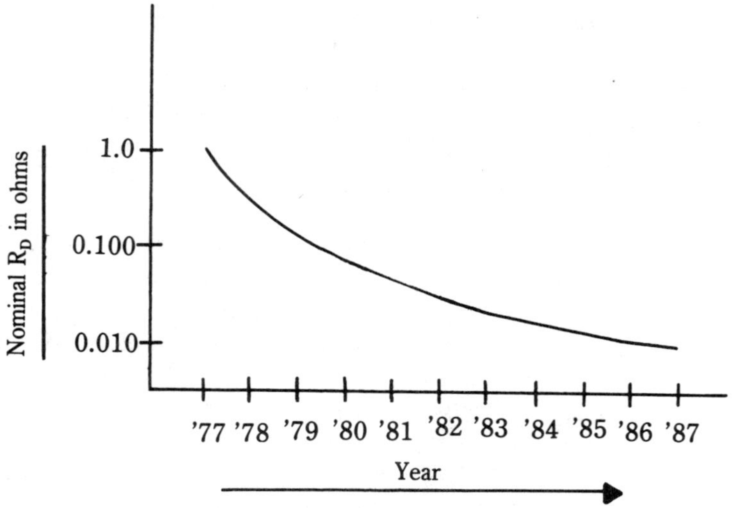

2-1 A decade of improvement in R_D of 50 – 100-volt power MOSFETs.

necessarily higher for power MOSFETs with higher voltage ratings; this follows from the fact that R_D is proportional to the 2.5 power of the voltage rating. This translates into readily available 500 V devices with R_D in the 0.5 ohm to 2.5 ohm range during the 1987 to 88 period. Although high compared to the 50 to 100 V units, the high voltage MOSFETs have nevertheless improved via a similar learning curve to that depicted in Fig. 2-1.

Shown in Fig. 2-2A is the partial schematic diagram of a switched power MOSFET operated at high frequency and high voltage, and often with a partially inductive load. Under such circumstances, the internal body "diode," actually, the collector-base section of a parasitic transistor, does not function well as a free-wheeling diode to provide a path for the reverse current needed by such a load. Therefore, an external fast-recovery diode is used for this purpose. However, at high voltages, the forward-voltage drop of the external diode is not low enough to prevent current from passing through the body diode. That is why the series-connected Schottky diode is also employed—it blocks any reverse current from passing through the internal body diode. Thus, virtually all reverse current is diverted through the externally connected fast-recovery diode. That is why the internal body diode is depicted in dotted lines, it is effectively deactivated. This is a clever approach, to be sure. However, the Schottky diode incurs losses, and the external fast-recovery diode is hampered in its action by the inductance of the connecting leads. Even worse, the parasitic transistor we refer to as a diode might be vulnerable to being turned on—usually a destructive occurrence.

A better performing and more cost-effective implementation is shown in Fig. 2-2B. Here, an improved power MOSFET with an internal fast-recovery diode is used. The internal body diode is again depicted in dotted lines, for it is electrically inactive to the reverse current; it is not allowed to function as the

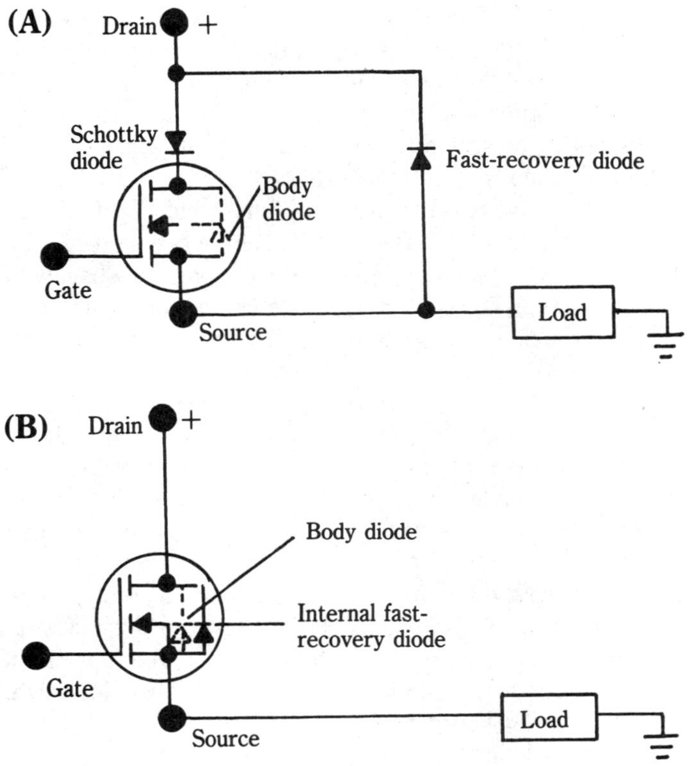

2-2 Circuit simplification via MOSFET with internal fast-recovery diode.

free-wheeling diode of the circuit. This function is taken over by the internal fast-recovery diode that has a lower forward-voltage drop than does the body diode. The designer must specify this type of power MOSFET. It is a specialty item in the sense that most of these devices will not have an internal fast-recovery diode. This is because the body diode suffices as a free-wheeling diode for a great many applications. Also, many power-supply circuits do not need a free-wheeling diode.

The domain of the "20 kHz switcher" has been both long and extensive. During much of this plateau in power-supply technology, the basic idea was to obtain good bipolar switching transistors—those with minimal conduction and switching losses. Although such switch-mode power supplies actually ventured into the 30 kHz to 40 kHz region, operation at higher switching rates tended to run into diminishing returns because of mounting losses in these power transistors. However, the manufacturers of bipolar transistors must have read the ads of the power MOSFET companies, which delighted in announcing the impending doom of the bipolar power transistor for switch-mode power supplies. For, soon enough, 100-kHz switching bipolar transistors became mundanely available, and for more exotic cost, you could obtain such devices with 200-kHz switching-rate capability.

Motorola and TRW were notable pioneers in the development of these "quantum-jump" switching transistors, featuring dramatic improvements over the "20-kHz types." Not only were the improved devices capable of much greater switching rates, but they became available at voltage ratings enabling off-the-line design of switch-mode supplies. Representative of such bipolar power transistors are the Motorola MJ16004 and the TRW 2N6583. Both can be fed from a full-wave rectifier working from a nominal 230 V line.

Not to be outdone, SGS-Thomson has developed a so-called *easy to drive* (ETD) bipolar power transistor with a 1000 V rating. The boasted reduction in drive power comes about not via a high current-gain, or β parameter, but rather by a drastic reduction in stored charge. This is achieved largely through an improved fabrication process in which the degree of digitation of the emitter fingers is carried out to a higher degree than was feasible in previous transistors. One such transistor bears the designation BUF410A. At 100 kHz, the switching losses in this transistor are inordinately low, making possible efficient off-line PWM regulation into a 300 watt load from a single BUF410A.

A simple test technique has come into prominence for evaluating the efficiency of various devices as power switches in switch-mode power supplies. Rather than deal with the intricate details of rise time, storage time, fall time, and conductive behavior, the overall effect of these many and variable dissipative losses can be ascertained quickly by monitoring the temperature rise during operation under prescribed line and load conditions. Not only does such a test summarize switching, conductive, and drive losses in the device under test, but it also includes the heat removal situation. Thus, a device with low thermal resistance between package and heatsink will, other factors being equal, show up better than a similar device with inferior package design. Although this test procedure circumvents numerous calculations, it provides an overall figure of merit geared to switching performance in a real-world situation.

Designers have been intrigued by the salient features of the power MOSFET. These salient features have included freedom from secondary breakdown, high frequency capability, easy drive conditions, ease of paralleling, excellent thermal stability, and sometimes, the use of the internal body diode as a free-wheeling diode. Also, the power MOSFET has become cost-effective, and is available in a wide range of current and voltage ratings. At the inception of this device, only n-channel types were commercially available and the perception tends to linger that this remains the case.

Although it probably remains true that the bulk of developmental effort has been expended for n-channel MOSFETs, most of the major vendors now supply p-channel MOSFETs also. These MOSFETs have respectable voltage and current ratings and for many practical purposes can be considered as matches for certain n-channel models. The initial incentive to develop p-channel units was for complementary symmetry circuits in the power stages of audio and stereo equipment. Later, p-channel MOSFETs also were optimized for use in switch-mode power supplies and in inverters. Supertex Inc. and International Rectifier Corp. were two of the pioneers in p-channel MOSFET technology. Often it is found that designs predicated around p-channel MOSFETs yield more straightforward solutions to circuit problems in certain systems. This closely parallels the situation with pnp bipolar transistors, which also lagged behind npn types in historical development.

Single-packaged arrays of power MOSFETs can be particularly convenient when both n-channel and p-channel devices are incorporated. Thus, the Supertex VQ3001 and VQ7254 families contain closely matched n-channel and p-channel MOSFETs. There are four devices in the 14-pin package, two of each type. Optimal circuit flexibility is provided because each of the four devices is an individual MOSFET with electrically independent terminals. These arrays merit consideration as complementary drivers for MOSFET switchers operating at higher power levels.

In general, if ultimate performance parameters are sought, p-channel MOSFETs are not likely to stack up as well as their n-channel counterparts. For example, if the very lowest R_D is required, time is better spent searching among n- than p-channel devices. The reason the two types aren't exactly equivalent is because the resistivity of p-type silicon is greater than that of n-type silicon. Thus, where matched R_D's are encountered, the p-channel device will utilize a larger die. This, in turn tends to increase input capacitance, and affects other parameters, including cost. Nonetheless, there are numerous applications where circuit performance can be about the same for both types.

Power MOSFETs for logic-level drive

It has long been known that various circuits could be devised in which a power MOSFET could be actuated by a logic-level signal. For optimum results, it has been necessary to select a power MOSFET with a low threshold voltage, and preferably a high transconductance. Even so, such circuits operated in the linear region of the power MOSFET's output characteristics. In other words, the power MOSFET could not be driven into its saturation region.

RCA has pioneered power MOSFETs with very low threshold voltages—on the order of one or two volts. These devices, known as logic-level DMOS FETs, turn on hard at five volts applied between gate and source. Such devices are not, as was previously the case, limited to small signal applications; early models already featured multiampere and several hundred volt ratings. Second sourcing took place by other semiconductor firms and the trend has been to increase load handling capability. At the same time, various trade offs have accompanied the basic low-threshold characteristic; for example, Supertex has introduced such a low-threshold device with a 10 ampere, 100 V rating that has inordinately low input capacitance—about 350 pF.

The developmental effort of logic-level thresholds has predominantly involved n-channel power MOSFETs. It will be interesting to see whether this will be followed by low-threshold p-channel types. Although conceivably useful to the circuit designer, there are inherent difficulties in producing such types, not the least of which is cost. This has to do with contradictions between low R_D and low input capacitance as functions of die size, which in turn, bears heavily on yield. However, practical compromises have been made in higher threshold p-channel MOSFETs that have been made to approximate the performance of n-channel units in complementary symmetry applications. If manufacturers feel there is sufficient demand for low-threshold p-channel devices, it is likely that satisfactory units will become available.

Bipolar transistors vs power MOSFETs

There has been a good measure of confusion concerning the relative merits of bipolar transistors and power MOSFETs for switching power supplies. Some of this has undoubtedly been because of the nature of advertising literature. In particular, it has been all too easy to draw the implication that bipolar transistors were on the way to obsolescence. This appeared to be the case at times, but surprising improvements in the switching capability of bipolar transistors has kept them in the race as competitive devices in certain application areas, and as superior devices in others.

Generally, it is more profitable to use bipolar transistors for switchers operating in the 20 kHz to 40 kHz region where the preponderance of switching supplies have been designed for more than a decade. This is because the sum of conductive and switching losses tends to be less in bipolar transistors at low switching rates than the power MOSFET. This is especially true for power MOSFETs operating at several hundred volts and higher. Conversely, high frequency switchers—those operating at several hundred kHz and higher—tend to be more efficient when designed with power MOSFETs, for in this portion of the spectrum the overall losses in the power MOSFET tend to be lower than is readily obtained from bipolar transistors.

For switchers operating in the 40-kHz to 200-kHz region, it sometimes is not too clear which device should be selected. If cost is not too important, you

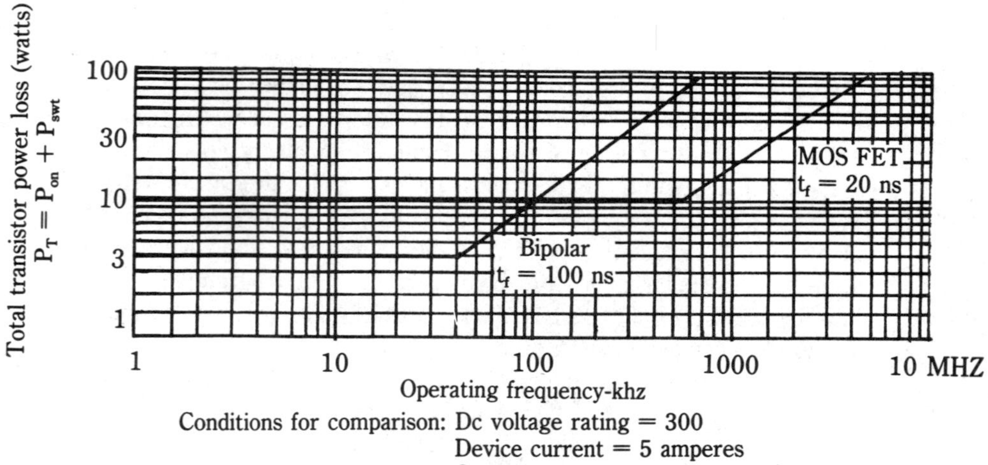

Conditions for comparison: Dc voltage rating = 300
Device current = 5 amperes
Operating temperature = 100°C
Circuit. . . . PWM regulator

2-3 Typical comparison between bipolar transistor and power MOSFET.

can select bipolar transistors with inordinately good high frequency performance. But, there are also power MOSFETs with inordinately low conductive losses (low R_D). In such instances, much depends upon the operating voltage, for the R_D of power MOSFETs tends to increase as the 2.5 power of rated voltage. In any event, an evaluation of the two devices should be made for devices that have appropriate voltage and current ratings for the actual application. Otherwise, the comparison will be between apples and oranges, as it were.

The performance curves of Fig. 2-3 exemplify what has been said in this regard. Here, the overall switching losses of 300-V, 5-ampere devices are compared under realistic conditions where both are assumed to operate with junction temperatures of 100°C. For the purpose at hand, consider the output section of the MOSFET to be a "junction"—a common reference found in the electronic literature. Note the breakpoints in the curves. This is where switching and conductive losses are equal. What counts in establishing losses as switching devices is the overall losses, that is, the sum of switching and conductive losses.

Of course, switching efficiency, although the main consideration, might not be the only factor entering into evaluation. You usually must also look at cost, reliability, ease of drive, and availability from vendors, i.e., second sourcing. Reliability can be influenced by design, because an appropriate circuit properly laid out can go a long way in minimizing switching transients and peak currents. It is true that the power MOSFET is not susceptible to the secondary breakdown that is often the nemesis of bipolar transistors. And, neither is the power MOSFET likely to commit suicide via thermal runaway. But, it too, will be destroyed by gross overloads. Although commonly likened to the electron tube in certain of its characteristics, it is by no means as forgiving as the tube. In particular, the power MOSFET is vulnerable to transients finding their way to the gate circuit. Some kind of gate circuit protection, such as a Zener diode or an MOV varistor often pays big dividends in operational reliability. Also, as is well known, the gate might be damaged or destroyed by static electricity imparted during handling or construction.

It is evident that each device tends to have its domain of superiority. It is also true that a considerable applications overlap exists where one or the other device might be effectively used. Above all, it should be appreciated that both devices have given good accounts of themselves in the field. Both devices have sufficient maturity so that consistent results and reproducible operation should be forthcoming where reasonable care has been taken to prevent violation of ratings. In practice this generally boils down to over voltage and over current protection, and to due consideration given to transients and parasitic oscillation.

Developments leading to the IGBT power switch

As with biological species, man's creations can be seen to follow evolutionary development. Consider, for example, the *insulated gate bipolar transistor*, usually referred to as an IGBT. This relatively new power switch owes its existence to both MOS and bipolar technology. Indeed, its main attraction is that it combines in a single monolithic structure some of the most desirable features of MOSFETs and bipolar power transistors. For example, it exhibits the low-drive input characteristics of the MOSFET, together with the low dissipation of the collector circuit of the bipolar transistor. This can be very meaningful in terms of operating efficiency in the several hundred volt region and higher where the power MOSFET develops excessive drain resistance, R_D.

In the early days of op amps, a smart thing to do was to use discrete elements in an amplifier having an FET input stage, and an output stage comprising bipolar transistors. You then acquired the desirable combination of high-impedance input together with low-impedance output. This scheme was later refined in the form of a single monolithic structure, the BIFET op amp. Although such op amps were low power level devices, the seeds were sown for somewhat similar devices with appreciable power handling capability. Paradoxically, the semiconductor architecture of the power device (IGBT) turned out to be simpler than that of the FET input op amp.

Another arrangement in the evolutionary path leading to the IGBT was the BIMOS power switch. This was actually a hybrid circuit developed and optimized by the International Rectifier Corp; it comprised a MOSFET input stage driving a common-base configured bipolar power transistor. The topography of such an amplifier identifies it as a cascode circuit. The overall advantage here is that the common-base connection allows a bipolar transistor to operate at an appreciably higher voltage than pertains to the conventionally used common emitter circuit. Yet, the inordinately low input impedance of the common-base circuit is of no applications consequence, being substituted by the very high input impedance of the gate circuit of the MOSFET input stage. Unfortunately, this scheme did not lend itself to reliable and cost-effective monolithic fabrication.

The eventual outcome of these FET-bipolar combinations was the IGBT, now a formidable contender for use in many power-supply applications.

An IGBT by any other name

The IGBT often is called various names by different manufacturers in their endeavors to call attention to proprietary fabrication processes. This is not a new situation, because power MOSFETs are referred to as HEXFETs (International Rectifier Corp.), T-MOS transistors (Motorola), and SIPMOS FETs (Siemens). Some alternate names and abbreviations of the IGBT are Comfet (original RCA device), IGT for Insulated Gate Transistor, GEMFET for Gain-Enhanced MosFET (Motorola), and MOSIGBTs (IXYS Corp.). This list does not exhaust the nomenclature for these two power devices, but suffices to show that there has been less standardization than exists with bipolar power transistors, which are almost universally referred to as such.

Whereas symbols for power MOSFETs and for bipolar power transistors have long enjoyed substantial standardization in schematic diagrams, you must

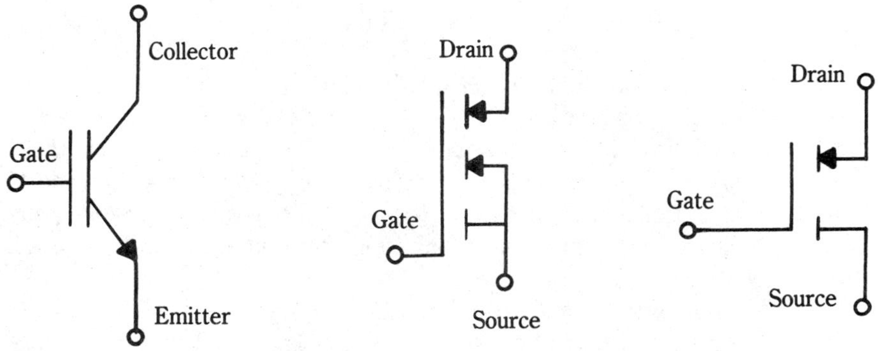

2-4 Some symbols depicting the insulated gate bipolar transistor.

be prepared to encounter a variety of symbols for the IGBT. Three of these are shown in Fig. 2-4. Note the common idea depicted is the isolated or insulated input electrode, the gate. In differing ways the illustrators have tried to convey the inclusion of a pn diode in the output circuit. This, of course, is implied in the symbol representing the output circuit as the collector and emitter of a bipolar transistor. Somewhat more true to the operational physics of the IGBT are those symbols showing a forward-biased diode in the ''collector'' lead. The for-

2-5 Equivalent circuits of the IGBT.

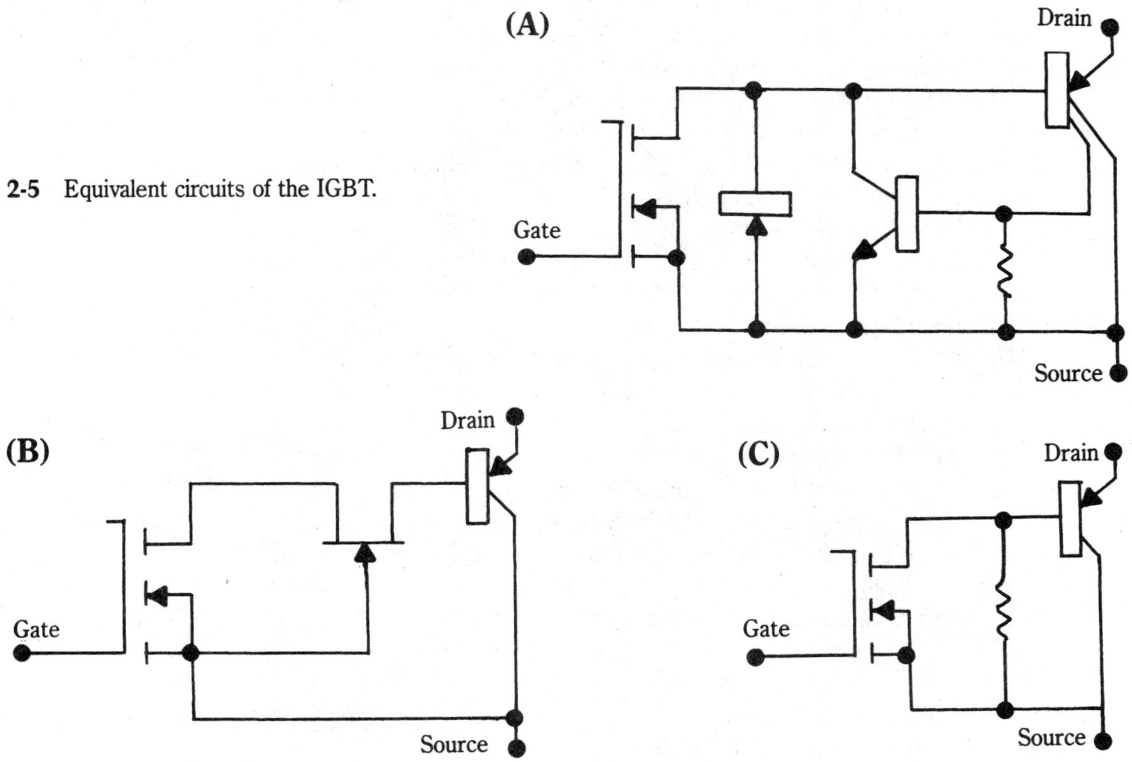

ward-biased diode can be interpreted as the emitter-base section of a pnp transistor in a common-collector circuit.

It appears that the center symbol, while not the quickest nor easiest to draw, best depicts the internal nature of the IGBT. In any event, all three of these symbols indicate a device analogous in polarity to an n-channel MOSFET or an npn bipolar transistor. Such a device has received the preponderance of development. Forthcoming p-channel or pnp units would be symbolically depicted similarly to the three diagrams of Fig. 2-4, but with the sense of all arrows reversed.

Simplified equivalent circuits of the IGBT are shown in Fig. 2-5. Circuit 2-5a is interesting in that you see the possibility of latch up from the SCR-like connection of the two transistor structures. Indeed, special processing techniques are needed to prevent such latch up. Circuit 2-5b represents another way of interpreting the multilayered structure. Circuit 2-5c suggests the behavior of the IGBT from the circuit designer's viewpoint. Here, you can deduce the MOS input characteristic, current amplification in a pnp common-collector transistor, and the pn diode offset from the base emitter section of this transistor.

The structure of the IGBT

It is evident that the IGBT appears at its input as a MOSFET, but at its output much as a bipolar transistor. Is it, then, a bipolar transistor with a modified base-emitter junction, or is it a power MOSFET with a modified drain structure? From the fabrication viewpoint, it turns out that the latter situation best describes this device. Indeed, it is essentially a power MOSFET in which the doping of the drain region of the substrate produces a pn junction, reversed however from the base-collector junction of an npn bipolar device. As is characteristic of pn junctions, conduction is assisted by minority carriers. Thus the IGBT is no longer a simple majority carrier device, as is the case with conventional MOSFETs. A beneficial result is that the conductivity developed in the substrate region can be greatly enhanced, thus reducing R_D. This, especially for high-voltage units, is appreciably less than R_D of the ordinary power MOSFET. Thus, there is less internal power dissipation during on times when the device is used as a switch. The structural comparison between basic MOSFET and IGBT devices is shown in Fig. 2-6.

As first marketed, the IGBT was soon found to exhibit a rather discouraging behavior—its tendency to latch up from over current, or from high temperature operation. In essence, the whole structure took on the characteristics of a four-layer diode. However, via appropriate doping profiles and structural arrangements, the latch up tendency has been largely overcome. Put into practical terms, by the time an improved IGBT can be provoked into the latch up mode through overloading, both MOSFETs and bipolar transistors probably would have been destroyed. The fact that the IGBT has gained a foothold as a motor-drive device bears witness to this, because motors are notorious for their transient current and power demands.

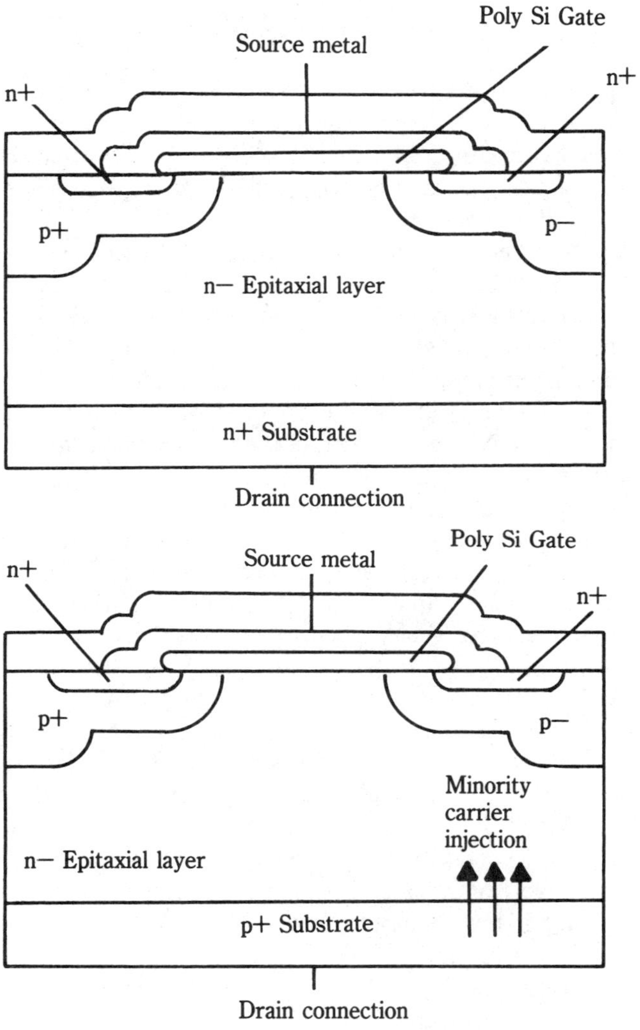

2-6 The power MOSFET and the IGBT.

The alluded pn junction in the output of the IGBT can give rise to some confusion regarding the operating physics of the device. Because, in an npn bipolar transistor, the "output diode" (the base-collector pn junction) is reverse biased in normal circuit operation. In contrast, the "output diode" (the n epitaxial layer p^+ substrate pn junction) of the IGBT is forward biased in normal circuit operation. Surely, the simplistic view of the IGBT as a tandem MOS-bipolar device requires some qualification. At fault is the alluded npn "output transistor," which unfortunately appears in some IGBT symbols. It is more accurate to view the "output transistor" as a pnp-type. The drain terminal of the IGBT then derives from the emitter of this pnp "transistor."

The basic behavior of the IGBT

The detailed application intricacies and the performance characteristics of any power device is, of course, best gleaned through deep study of the engineering specifications and the applications notes provided by the manufacturers of the device. For practical guidance purposes, however, the following outline of the behavior of the IGBT will prove useful:

The input: at the input, the IGBT is similar to what the driver "sees" at the gate of a power MOSFET—the gate is, basically, a capacitor. A difference is that the apparent capacitance of the gate-source circuit of the IGBT can be considerably less than that presented by the input of the power MOSFET. The primary reason for this is that the influence of drain voltage on the gate is much less in the IGBT than in the MOSFET. In other words, there is less Miller effect in the IGBT. You are reminded of the screen grid or tetrode versus the triode electron tube, although the analogies in operating mechanisms in the tubes and the solid state devices are not strictly parallel.

The output: marketing literature often depicts the IGBT as a marriage between MOS and bipolar technology. In a certain sense, this is true. However, you soon learn that the output characteristics of the IGBT differ in nontrivial respects from what you would expect from the collector-emitter circuit of a bipolar device.

Specifically, the drain source section of the IGBT incorporates a forward-biased pn diode; this is in marked contrast to the reverse-biased collector base pn junction of the bipolar transistor. This being the case, you have to put up with a conduction "gap" in the IGBT—no conduction exists in the drain circuit below about 0.7 volt. Inasmuch as the IGBT is usually implemented as a high-voltage device, this conduction offset tends to be of negligible consequence.

The real reason for likening the output circuit of the IGBT to that of the bipolar transistor is that minority carriers are involved in both devices. This results in much greater current densities in the IGBT than in the MOSFET, where only majority carriers cause current in the output circuit. The "side effect" or penalty incurred in the IGBT is greatly decreased frequency response. This stems from the storage phenomenon associated with the minority carriers; it takes time for the minority carriers to recombine with majority carriers and be extinguished. This is manifested by long fall times in the switched waveform. But, this is the very mechanism that ultimately limits fall time in bipolar transistors. So, the advertisements of IGBTs are not altogether off base after all.

The overall behavior of the IGBT is essentially a high-voltage, high-current, relatively slow-speed switching device. Of course, it can be applied in linear circuits too, where its low drive and low internal dissipation at maximum conduction can advantageously be used. Although, voltage and current ratings in the vicinity of 1000 volts and 30 amperes are common, switching rates are generally maximized in the 5 kHz to 20 Hz region. However, the penalty paid for using IGBTs with higher switching-rate capability is higher ON resistance (R_D). It goes without saying that IGBTs intended for 60 Hz or 400 Hz motor control can offer formidable competition to previous devices used in this service, i.e., SCRs, triacs, GTOs, bipolar and darlington transistors, and power MOSFETs.

Like the power MOSFET, the IGBT is resistant to secondary breakdown or similar destruction from concentrated energy effects. A nice feature of the IGBT is that its output resistance (R_D) increases relatively slightly with temperature compared to a power MOSFET.

Costs: the IGBT structure is so similar to the power MOSFET, that very little extra expense is involved in its fabrication. When compared with similarly rated devices, it has been cost competitive from its inception; there is little need to await cost benefits from a prolonged learning curve as had been the marketing experience of SCRs. Implementation is straightforward—no commutating circuitry or exotic drive requirements are involved. Snubber circuits for protecting the safe operating area are often desirable, but the situation is probably more relaxed than with bipolar devices, where secondary breakdown is an important factor.

Diode: in the IGBT, the parasitic diode that appears between drain and source in MOSFETs, does not manifest itself. This means that an external free-wheeling diode is needed in some applications, where it represents a cost item. However, this is in many instances a blessing in disguise, because designers have often found that the parasitic diode of the MOSFET output circuit does not always yield optimum performance or protection. With an external free-

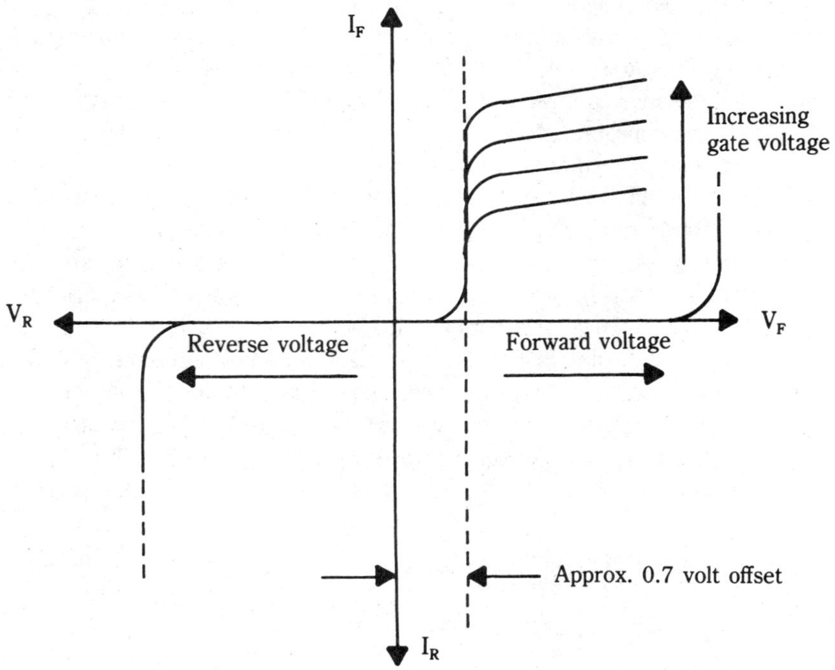

2-7 The output characteristics of the IGBT.

wheeling diode, you can tailor the voltage, current, and reverse-recovery characteristics of the diode for optimum overall results in the circuit.

Output characteristics: the output characteristics of the IGBT are shown in Fig. 2-7. The drain-current vs. drain-voltage curves resemble those of the power MOSFET, except for the slight voltage offset (no conduction below approx. 0.7 volt). Note also the reverse-blocking feature. This, however, can vary greatly according to various performance trade offs incorporated by the manufacturer.

When should the IGBT be used?

The IGBT shines at high voltages and at high currents, but cannot compete with bipolar or MOSFET devices where high switching rates are needed. IGBTs can still maintain advantages of low conductive losses in the 20 kHz region, but begins to display appreciable switching losses at much higher frequencies than this. Figure 2-8 shows on a comparative basis where the IGBT merits consideration with bipolar transistors and power MOSFETs. It is readily seen that the IGBT does not compare favorably (using forward voltage drop as criteria) with the other devices at low (100-volt) operating voltage.

However, at a higher operating voltage—600 volts—the IGBT surpasses its rivals in both the matters of lower voltage drop (R_D) and current capability. The superiority of the IGBT at yet higher operating voltages becomes even more dramatic.

Of course, you might choose the IGBT for a relatively low-voltage application where its easy-drive feature might give it preference over a bipolar transistor, or where its reverse-blocking feature might give it preference over a power MOSFET. (Be sure to investigate the reverse-blocking characteristics where this behavior is important. Not all IGBTs have high capability in this regard.)

The crossover frequency, or that frequency region where IGBTs and the other devices exhibit roughly similar conductive losses depends considerably upon the parameters that have been optimized in the devices, but tends to be in the 250 to 350 volt region. At much lower voltages, the IGBT becomes less competitive; at much higher voltages, the IGBT tends toward superiority as long as its frequency capability is not exceeded.

Thyristors were not included in Fig. 2-8 because of their commutation problems when used in inverter or converter circuits operating from DC. Also, for simplicity, darlington transistors were not considered—their forward voltage drops being higher than single bipolar transistors, they would not alter the basic information conveyed by these device comparisons.

High speed or high reverse voltage blocking

The IGBT, as you have seen, is a modified power MOSFET. By introducing minority carrier injection into its drain circuit, much greater current density is

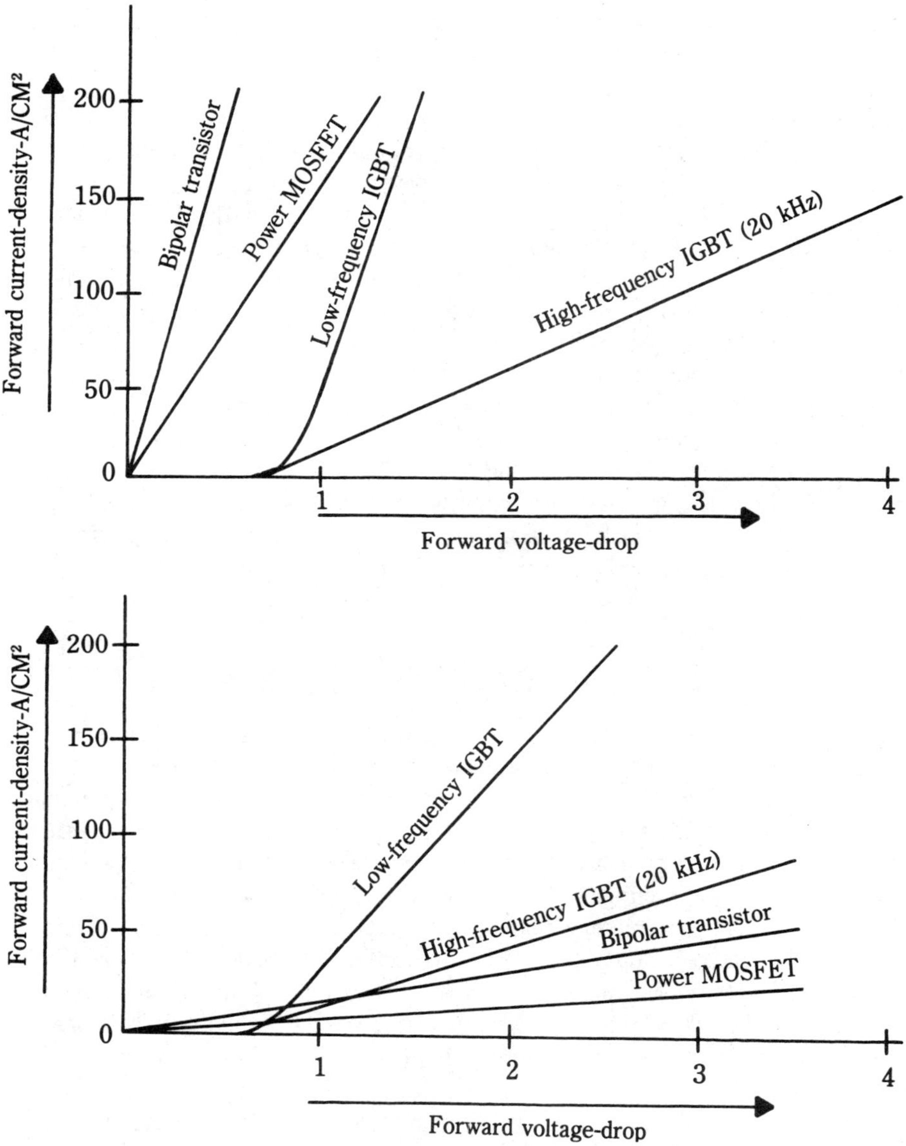

2-8 Forward voltage drop of several competitive power devices.

achieved than is possible in the original MOSFET structure. This technique, known as conductivity modulation, has yielded a power device as superior over the "plain" power MOSFET in some respects, because the power MOSFET excels the bipolar transistor, again, in some respects. The gist of these statements is that enhancement of one or several performance parameters often occurs at the expense of one or more other ratings. This is the familiar concept of trade offs—optimization of an operating parameter generally causes degradation of some other parameter(s). Relevant to the IGBT, the attainment of high-current and high-voltage operation and low R_D has been very much at the expense of speed. The culprit is storage phenomenon of the injected minority carriers.

One way to combat this situation is to employ processing and structural techniques that reduce the lifetime of the minority carriers. A popular method is to insert a buffer layer of n^+ material between the p^+ substrate and the epitaxial region of the device. This enables IGBTs to function fairly efficiently into the 20 kHz switching region, making them useful for power supplies and motor control where audible noise is not desired. As we are led to suspect, this buffer layer can be expected to produce side-effects.

Whereas the simple IGBT exhibits nearly symmetrical voltage-blocking behavior, the addition of the buffer layer greatly reduces the reverse voltage-blocking capability of the device. In many applications, this is not detrimental. Indeed, the unaltered power MOSFET also lacks any appreciable blocking capability for reverse voltage. In using the IGBT, you must determine whether the application requires reverse-blocking capability before specifying the device—manufacturers make both types. The most difficult situation facing a designer is where both high-frequency response and the ability to hold off high reverse voltage are needed.

Evolution of the MCT thyristor

The SCR is truly a mature semiconductor power device. Although it has continually been improved since its inception, it remains a dilemma in the design of power circuitry. The SCR has long been "king of the mountain" when it comes to handling high voltage and high current simultaneously, but its very nature deprives it of the controllability of other power devices. Once turned on, the gate or control electrode loses power to turn it off. Turn off only can be achieved by commutation techniques, that is by depriving the basic anode–cathode structure of operating voltage (and therefore, current). Thus, the SCR is much more easily implemented in AC than in DC systems. Commutation in an AC circuit occurs with the repetitive zero crossing of the applied voltage. Commutation in a DC system requires momentary interruption, or shorting, of the applied voltage. In light of this situation, considerable effort has been expended in search of some method of finding reliable and economical turn-off techniques that would essentially preserve the power-handling capability and the efficiency of the SCR.

One result of this quest has been the gate-turnoff thyristor, the GTO. It is essentially an SCR in which the regenerative feedback loop within the device can be momentarily broken by application of a negative pulse to the gate. The shortcoming of the GTO is the relatively high power required for such commutation—approximately one-third of the load current is needed. For this reason, designers have tended to favor other devices, such as power darlingtons where applicable. For utility power frequencies, and for heavy-duty applications, such

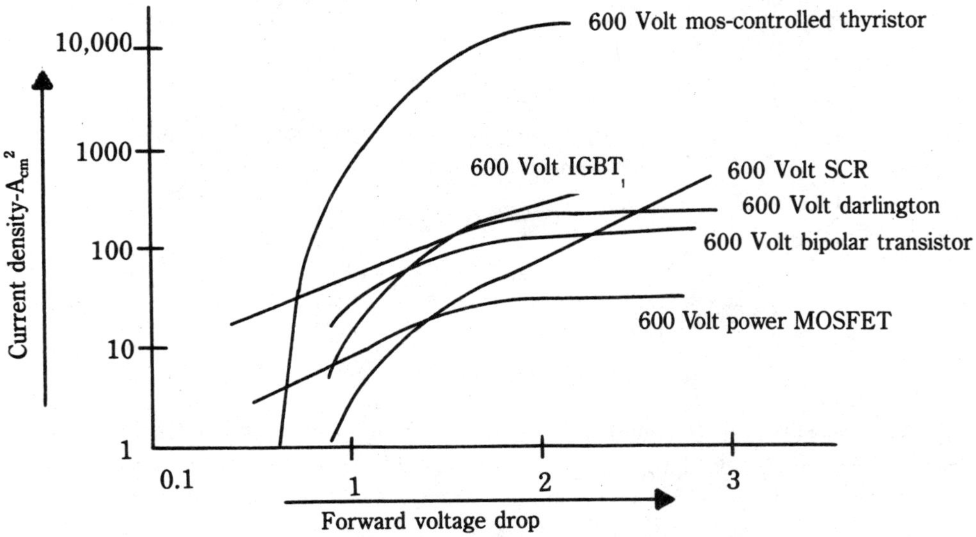

2-9 Voltage drop and current density of 600-volt power devices.

as the control of large traction and industrial motors, the GTO has proved itself a valuable power device.

A new power device has been developed that overcomes the high-current demand of the GTO gate, and yet displays output characteristics similar to that of the SCR. Known as the MOS-controlled thyristor, or MCT, a megawatt or more of load power can be turned off in about two microseconds via negligible gate power. Turn on times of 200 nanoseconds are readily achieved, again with negligible gate power compared to load power. More rapid turn off is thought feasible with fabrication improvements, or by deliberately allowing certain performance trade offs. Figure 2-9 shows the MCT to be in a class by itself with regard to high current capability and low forward-voltage drop.

The MOS-controlled thyristor

A proposed symbol for the MCT is shown in Fig. 2-10. Its salient feature is simplicity, an obviously desirable attribute for the engineer and the draftsman. Indeed, it would serve no useful purpose to try to depict fabrication details of this complex power IC. Nonetheless, the relevant basics of the device are revealed to the knowing eye. For example, the isolated nature of the gate is depicted in much the same manner as the gate electrodes of MOSFETs. You can glean from the symbol that the input to the device appears as a capacitance.

Also, the labeling of the electrodes—gate, anode, and cathode, is the same as that of SCRs. So, already you know that the device behaves somewhat as an SCR, but with an isolated or insulated control electrode. Somewhat more subtle

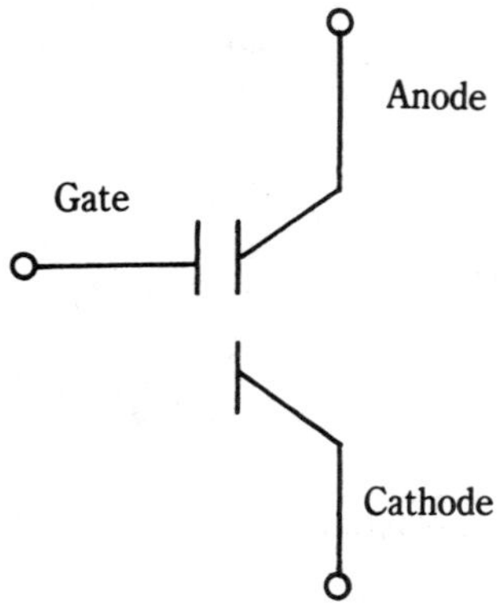

2-10 A suggested symbol for the MOS-controlled thyristor.

is the spacing between the righthand vertical segments—this is supposed to represent the voltage-blocking behavior of the npn transistor associated with the cathode. Inasmuch as the anode and cathode electrodes are slanted lines, the suggestion of bipolar transistors is created. Although appropriate arrows on these diagonal lines could further detail the npn and pnp transistors involved, the overall symbol, in the interest of simplicity, is more of a stylized than a pictorial view of the real thing. In any event, it should be helpful to correlate this symbol with the equivalent circuit shown in Fig. 2-11. It will be seen that the symbol, although somewhat strange at first inspection, is actually more natural than arbitrary in its depiction.

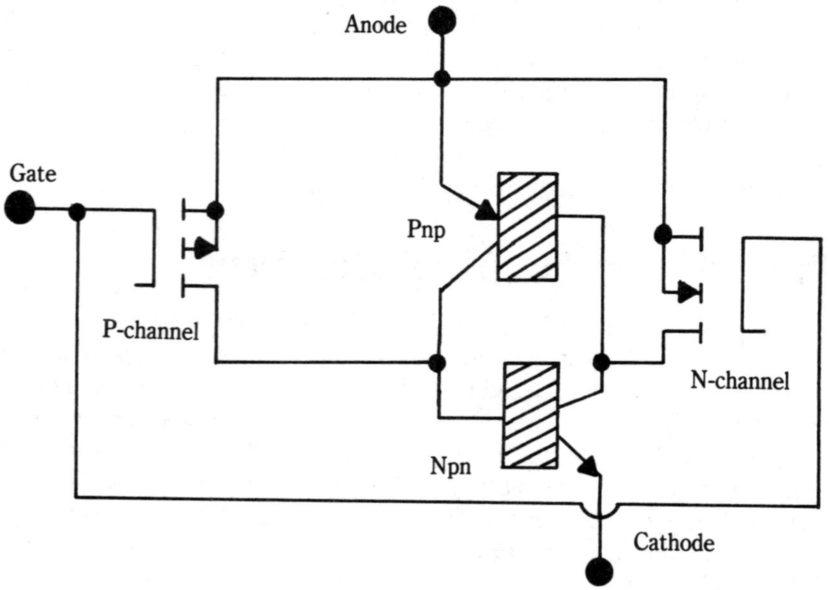

2-11 An equivalent circuit for the MCT.

Insights from the equivalent circuit of the MCT

The MCT is a complex IC and its structural details would be meaningful to semiconductor specialists, but to a much lesser degree to those whose primary interest is its circuit behavior. This can, in a qualitative, but useful way, be gleaned from an equivalent circuit, such as the one depicted in Fig. 2-11. Note that the two bipolar transistors are in an arrangement often used to simulate the behavior of an SCR. Thus, if a positive trigger pulse were applied to the base of the npn transistor, this transistor would begin to turn on; in so doing, it would deliver a turn-on bias to the pnp transistor, which in turn would reinforce the original turn-on signal at the base of the npn transistor. It can be seen that a regenerative action takes place with each transistor driving the other to saturation. This actually occurs in a very short time, both transistors having abruptly switched from their off states to hard conductivity in their on states.

This sequence of events simulates the triggering of an SCR. Once the two transistors have completed their regenerative action, removal or reversal of the trigger-pulse exerts no further control—this, of course, is the shortcoming of SCRs when used in DC circuits. Actually, some SCRs might, under some circumstances, turn off from an extremely large negative pulse. This, however, is of no practical significance because the required pulse power would tend to be comparable to the actual load power. Moreover, such levels of trigger power would damage the gate structure, if not immediately, certainly after limited use.

To circumvent this situation, the GTO thyristor was developed. It is an SCR in which internal structural arrangements have been made to alleviate the obstinacy of the locked-up condition. The result is that the GTO can be reliably turned off with a negative gate pulse about one third the load current. Although the GTO thereby became a practical device, the search continued for an SCR with less demand for turn-off power. The culmination of this search has been the MCT in which both turn-off and turn-on power are negligible compared to load power.

The equivalent circuit of Fig. 2-11 indicates why this is so—MOSFETs are interposed between the gate terminal and appropriate triggering points in the transistor pair forming the SCR proper. It is not hard to see the possibility of the overall circuit being receptive to both turn-on and turn-off pulses.

General operating conditions of the MCT

The salient feature of the MCT is its control characteristics. Like an SCR, it is either fully on or fully off. And, like a GTO, it can be turned on or off by applying a pulse of appropriate polarity to the gate. The pulses are applied to the gate with respect to the anode. A nominal plus or minus 12 volts suffice for such control, with the negative pulse for turn on, and the positive pulse for turn off.

There is a tendency for turn on to be initiated with less than 10 volts. Note that these gate-pulse polarities are opposite to those required in the GTO thyristor, where a positive gate pulse turns the device on. In any event, the current and power needed to control the MCT is much less than with the GTO thyristor. The input of the MCT appears as a capacitor of a value depending on the size of the device. For a 50-ampere-1000-volt unit, the input capacitance might be on the order of 7000 pF.

The speed of such a 50 ampere, 1000 volt MCT can be gleaned from the following parameters: Turn on is very fast—on the order of 200 nanoseconds, and is therefore of negligible consequence in limiting the switching rate of the device. Storage time can be around 500 nanoseconds, and actual turn off time in early models is about 2000 nanoseconds, i.e., 2 microseconds. In order that the input capacitance of the MCT will not limit the switching rate, the gate-control pulses should derive from a low-impedance source.

A notable feature of the MCT is its inherent electrical ruggedness. For example, a di/dt rating of 2000 can be achieved. This means that the device can be turned on into a short-circuited load, relying on protection from fuses or circuit breakers, rather than from fast-acting electronic circuits. At the same time, the MCT can withstand anode to cathode transients of 20,000 volts per microsecond without being forcibly turned on. Moreover, it appears that higher operating temperature will be feasible than with SCRs, providing the MCT is appropriately packaged. (TO-3 packaging can, for example, more advantageously exploit the MCT's thermal behavior than can be expected from plastic packages.)

Because of its ruggedness, the MCT can sometimes operate reliably in snubberless circuits, but like other devices, it, too, has its safe operating area (SOA). Exceeding the boundaries of the SOA can, as with other devices, cause destruction. In general, a snubber circuit should be used to ensure that the load line remains within the SOA.

The MOS thyristor, an improved SCR, but not quite an MCT

An earlier version of an SCR controlled by a field-effect input-stage was simply known as an MOS thyristor. It has the easy turn on characteristic of the MCT, and for the same reason the turn-on pulse is applied to the gate of an intrinsic MOSFET. However, unlike the MCT, there is no turn-off capability. As with conventional SCRs, once this device is triggered into its conductive state, it latches. Turn off only can be accomplished by removing the anode – cathode voltage, or by reducing the anode current below a crucial value. Other than this, a minor applications difference is that the gate pulse is applied with respect to the cathode of the device, rather than to the anode, as in the MCT.

Figure 2-12 shows the symbol and the approximate equivalent circuit of the MOS thyristor. It is implemented in circuits in similar manner to ordinary SCRs. Thus, in DC circuits the MOS thyristor must be commutated, that is, forcibly turned off. In the push-pull power circuit of Fig. 2-13, C2 is the commutating capacitor. It conveys negative pulses to the anodes of whichever MOS thyristor

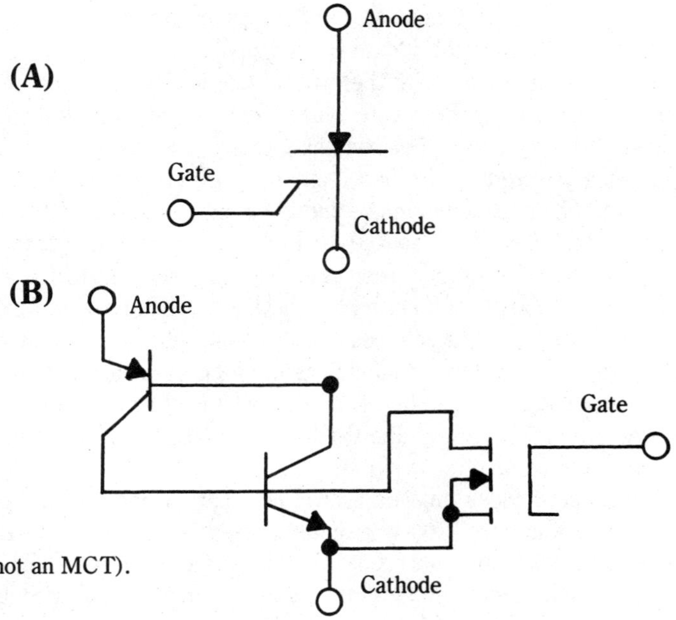

2-12 The MOS thyristor (not an MCT).

2-13 Driven inverter using two MOS thyristors.

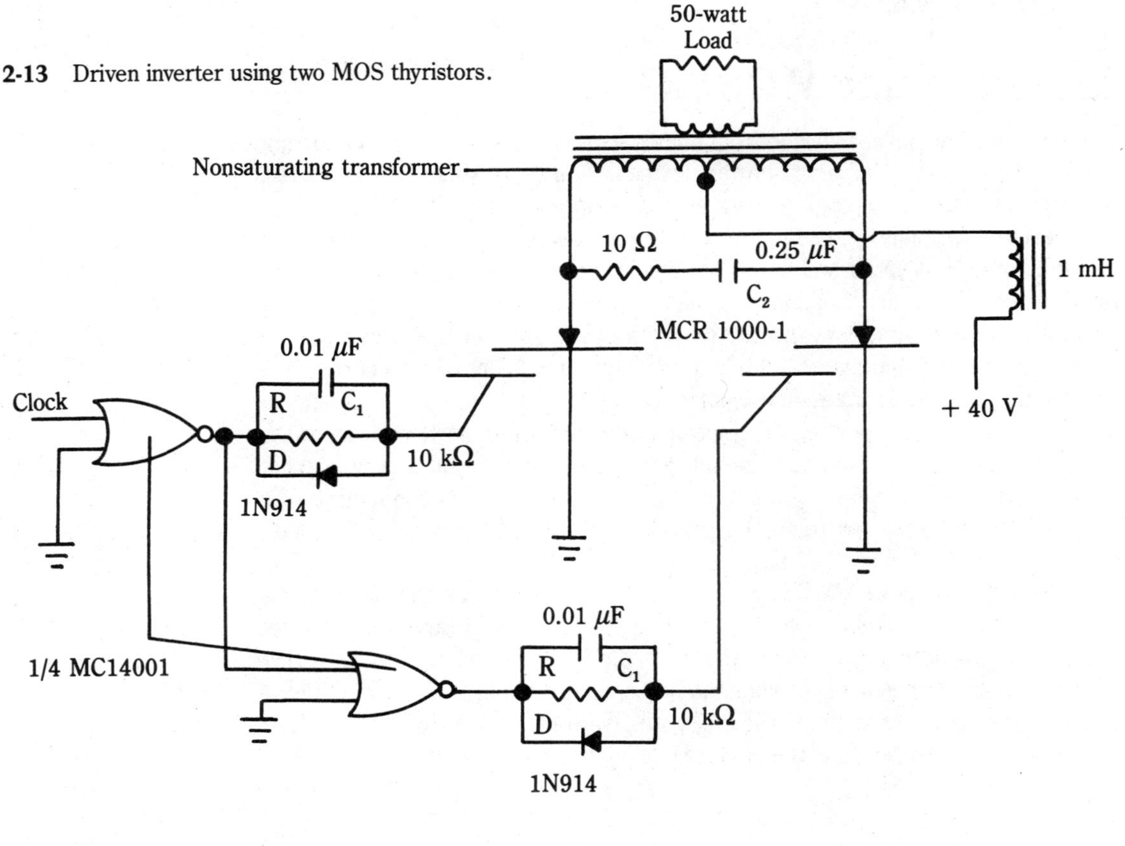

is to be turned off. These turn-off pulses are produced by the abrupt turn on of the alternate thyristor.

The salient feature of the drive circuit in Fig. 2-13 is to provide a low impedance so that thyristor gate charge can be initiated and dissipated speedily. Aside from making higher switching rates possible, the low-impedance drive is necessary to reduce the possibility of inadvertent turn-on of the thyristor from abrupt voltage changes in the anode circuit. Such premature turn on can result from a high-frequency source for the load, or from an inductive load.

In the drive circuit, resistance (R) prevents overdrive and capacitor (C_1) provides a speed-up function for the leading edge of the turn-on pulse. Diode (D) provides a low-impedance discharge path through the CMOS gate when this logic device goes low. Inasmuch as the logic gates are packaged as a quad, several of these gates can be used in parallel for even greater effect in minimizing delay time and in protecting the MOS thyristor against premature turn on from anode transients.

These devices enjoyed a brief life span during the mid eighties. Motorola made a 15-ampere, 200-volt model under the designation MCR 1000-4. Write ups in the trade journals created considerable enthusiasm. The probable reason for ending its production was the inception of the MCT, with its turn-off capability. Aside from the fact that you might encounter equipment using this device, the summarized discussion presented here is instructive, and some of it is applicable to the use of the MCT.

The bilateral MOSFET

The control of AC power at high frequencies generally has posed problems not readily answered by traditional power devices. The triac works fine at 60 Hz, but tends to turn on spontaneously at higher frequencies. 400-Hz units are commercially available; probably 1 kHz is about the limit of operation under best of conditions. SCRs can handle AC power in the 30 kHz to perhaps 50 kHz region; they, too, are hopeless at higher frequencies. Bipolar transistors (including darlingtons), conventional power MOSFETs, and the newer IGBTs can be made to exhibit bilateral conduction needed for AC power via certain circuit techniques. An example is a bipolar transistor connected to the conjugate terminals of a rectifier bridge. The bipolar transistor is not, of its own accord, a bilateral control device, and such circuit techniques are wasteful of power. What has been needed is a device that would provide the bilateral conduction of a triac, but at much higher frequencies, and preferably without the loss of gate control inherent in most thyristors.

The ordinary power MOSFET would perform as a controllable bilateral device if it could be freed from the shorting effect of its intrinsic body diode. Indeed, in synchronous-rectifier circuits, operation takes place with reverse-polarized drain-source voltage. Thus, an n-channel device in such a circuit delivers output current when the drain is negative with respect to the source. The trick employed to inactivate the body diode under such operation is to use a

MOSFET with a lower R_D than the nominal 0.7-volt threshold of the body diode. This technique is fine for this particular application is not readily applicable for general control of AC power where both halves of the AC cycle must be passed. It does, however, demonstrate that the power MOSFET would be "happy" to function as a bilateral control device if only the body diode could be taken out of the picture.

Consider the AC control circuit shown in Fig. 2-14. Assume the top MOSFET's drain is positive. This MOSFET will be subject to conventional gate control so that it will attempt to pass current to the load if switch SW is closed. But, how can this load current complete its path through the lower MOSFET? It will do so not by passing through the conductive channel of this MOSFET, but by passing through its body diode, which finds itself polarized to act as a shunt, bypassing the current around the channel. When the drain of the top MOSFET is negative, the two MOSFETs simply alternate roles. Over a complete AC cycle, load current is bilaterally passed in this fashion.

Note that neither MOSFET is required to operate as a bilateral device; rather, bilateral conduction of load current is the result of the two devices performing automatic switching functions with the aid of their body diodes. This action is different from that occurring in synchronous rectification, where operation is dependent upon low R_D of the MOSFETs.

From casual inspection of the symbol for the bilateral MOSFET, as shown in Fig. 2-15, it might appear that driving this device is a mere triviality—it only would be necessary to impress the appropriate gate-control signal between the gate and source terminals. When you see this symbol, it is only natural to think of

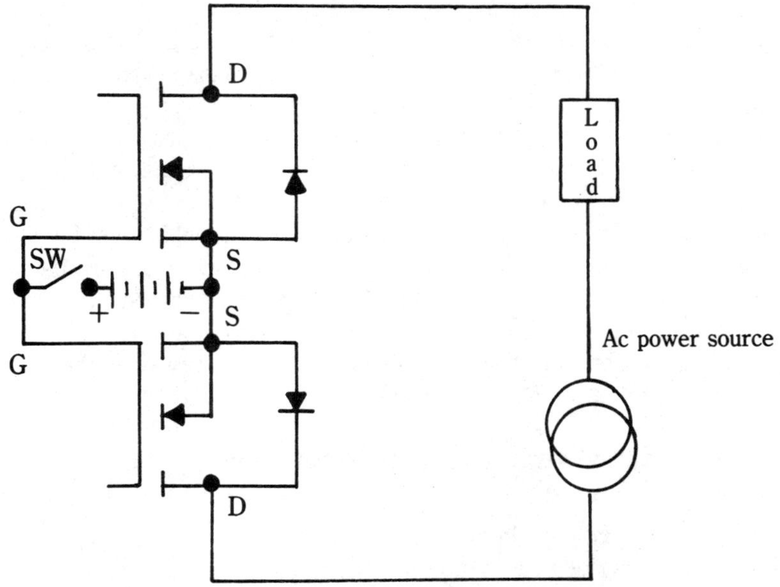

2-14 Circuit technique for achieving bilateral conduction by power MOSFETs.

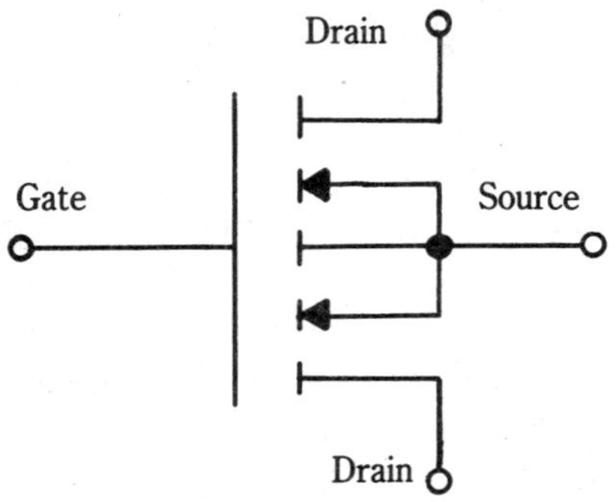

2-15 Symbol of bilateral MOSFET.

the source terminal as a "centertap" between the two drain terminals. As such, it would appear that no problems need occur from a directly coupled drive circuit; the source terminal, according to such reasoning, would be a neutral point with respect to the output circuit, a sort of "phantom ground," as it were.

The fallacy in such reasoning is that whereas the source terminal might be electrically suspended half way between the two drain terminals during OFF state of the device, things are quite different when ON voltage is applied to the gate. During a cycle of ON operation, the source terminal alternately swings one diode drop from one drain terminal to the other. This could lead to trouble in the drive circuit, where the unanticipated appearance of high voltage from the drain circuit could have devastating effects. What is needed, is some type of isolation in the gate-source circuit. This can be accomplished by a transformer, by capacitive coupling, or by an optoisolator. A nice way of driving the device, especially when sustained output is desired, is by means of an oscillator. The frequency of the oscillator should be much higher than that of the AC output circuit. Keying the oscillator off and on exerts the control. The high-frequency component in the output circuit is readily filtered out, if need be. In its simplest form, the oscillator would be capacitively coupled to the bilateral MOSFET.

An interesting aspect of this suggestion is that the oscillator could be amplitude modulated, making the bilateral MOSFET behave as a linear-control device of AC power. Such means of controlling AC power in loads portends certain advantages over triac or SCR control systems. It would be possible to work at much higher frequencies than with the thyristor schemes. Less EMI and RFI troubles would be encountered inasmuch as the chopped waveforms associated with thyristors would not be present. Also, although not always realized, very low power factors often attend thyristor systems; such would not be the case with linear control of the bilateral MOSFET.

Single-package complementary-symmetry MOSPOWER device

Circuit designers find complementary-symmetry devices intriguing; the "mirror-image" behavior of a pair of power transistors lends itself well to innovative techniques, and often results in reduced parts count. The complementary-symmetry concept first established itself with bipolar transistors because of the

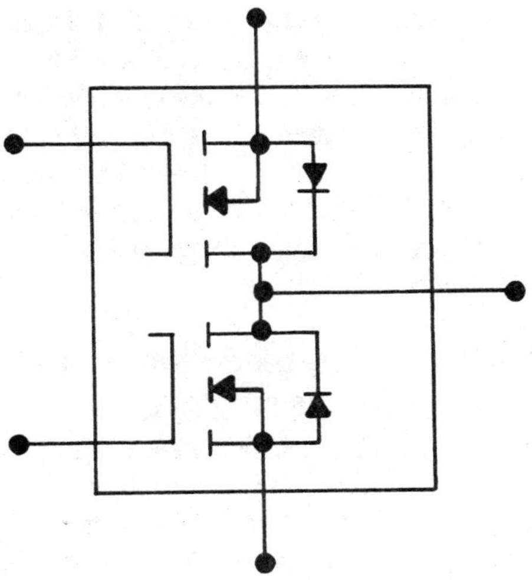

2-16 Symbol of single-package complementary-symmetry device.

availability of both npn and pnp devices. This was in contrast to the prior electron-tube technology in which operation with negative plate voltage was not possible.

With the advent of p-channel power MOSFETs, practical complementary-symmetry circuitry also came into prominence for MOSFET technology. For example, the push-pull output stage of audio amplifiers often has been considerably simplified through the use of an n-channel and a p-channel MOSFET instead of the conventional n-channel pair. In such equipment, the complementary-symmetry output stage can perform in push-pull fashion without the need for the traditional centertapped input and output transformers.

A recently introduced power device is comprised of an n-channel and a p-channel power MOSFET in a common package. Figure 2-16 shows such a device. Note that the drains of the two MOSFETs are internally connected. Inasmuch as this would naturally be the case in many complementary-symmetry circuits, the loss of application flexibility tends to be minimal. Siliconix Corp. has pioneered this device technology with its MPP500 family of complementary-MOSPOWER arrays.

Although simple in concept, there is more than initially meets the eye in these arrays. A problem that had to be met involved matching of the two units—it would be desirable that the n- and p-channel MOSFETs should have identical characteristics other than the polarity of their operating voltages. For the two MOSFETs to have the same R_D, the area of the die used in the p-channel device would have to be 2.4 times that of the n-channel device. This is because of the physics of MOSFET devices in which the majority carrier (electrons) of n-channel devices has greater mobility than the majority carrier (holes) of p-channel devices. This disparity means that a pair of complementary MOSFETs of equal power rating can be matched in R_D or in input capacitance, but not in both.

Up to this time designers of line frequency, and even audio frequency applications have tended to seek out discrete n- and p-channel MOSFETs with equal R_D. However, for the newer power-supply circuits working in the 100 kHz to several MHz region, it is often better to deal with equal input capacitances. Accordingly, the MPP500 family of complementary MOSPOWER devices comprise MOSFETs fabricated on equal area dies. In practice it usually turns out that fewer high frequency difficulties attend unequal R_D than unequal input capacitance.

The depletion-mode power MOSFET

The power MOSFETs long used in power-supply circuits have been enhancement types; these normally are off and require a certain forward-bias voltage to project operation into the conductive region. This has been true for both n and p-channel polarities. For most purposes, enhancement-mode operation works out well, but certain circuitries could benefit from a depletion-mode device; such a device normally is on and can be rendered less conductive by application of reverse bias. Depletion-mode devices have been available in the form of junction FETs, and were common in electron tubes. A shortcoming of the depletion-mode JFET is its relatively low power-handling capability.

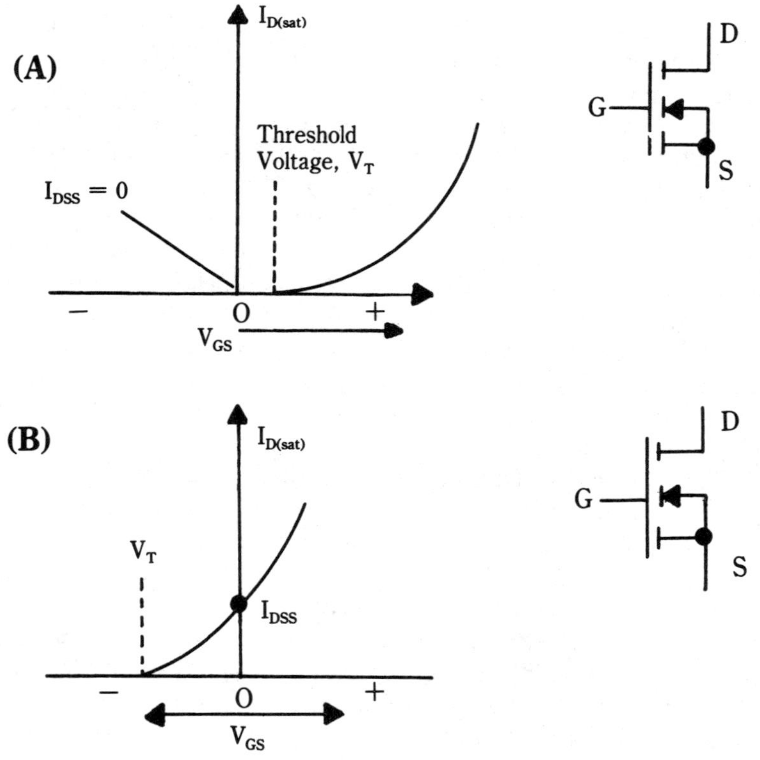

2-17 Comparison of depletion-mode and enhancement-mode MOSFETs.

The Siliconix Corporation has introduced a line of depletion-mode power MOSFETs. The essential difference between conventional enhancement-mode MOSFETs and the newer depletion-mode types is illustrated in Fig. 2-17. An interesting feature of the depletion-mode devices is that greater than zero-bias drain current results from forward-bias operation. In other words, operation can embrace both depletion and enhancement modes. This is reminiscent of electron-tube operation except that operation in the enhancement mode does not require gate current.

In the interest of simplification, the internal body diodes are not shown in the symbols of Fig. 2-17. However, these parasitic elements exist in much the same way in both depletion- and enhancement-mode MOSFETs. N-channel depletion MOSFETs have borne first fruits of developmental effort, but it is likely that p-channel depletion devices also will become available. Note the basic similarity of the transfer-curves of Fig. 2-17. It is as if the enhancement-mode curve was simply pushed over to the left to provide a new threshold voltage.

The simple two-terminal constant-current regulator shown in Fig. 2-18 derives directly from similar circuits used with JFET devices. However, much greater voltages and currents can be accommodated. If the Siliconix NOS100 is used in conjunction with a 1-kilohm variable resistor for R_S, constant-current operation from a 200-volt source can be obtained over a 5-mA to one-half ampere range. In general, an appropriate heat sink will have to be used, inasmuch as the MOSFET accomplishes its current regulation via internal dissipation, as in a rheostat.

Also, connecting the sources together and likewise, the gates, of two depletion-mode MOSFETs yields a bilateral AC switch that normally is on. A like connection of two enhancement MOSFETs results in a bilateral AC switch that normally is off. Thus, the functions of an electromechanical relay can be simulated, but at higher efficiency and higher switching rate, and in a maintenance-free fashion.

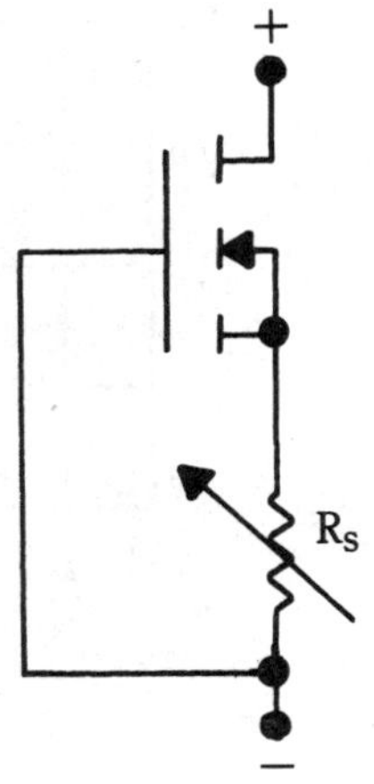

2-18 Two-terminal, constant-current regulator using depletion-mode MOSFET.

Refrigerated power devices

If a cause-and-effect relationship is known over a limited range of variables, it can be dangerous to predict behavior by extrapolation from a known trend or graph shape. This is a common error in economics, geology, physics, and other fields. An example is the phenomenon of superconductivity—knowing the way in which the resistance of some metals changes over a fairly wide temperature range still provides no clue of the "magical" disappearance of resistance at liquid helium temperatures. Similarly, scientists can't be sure that the "laws of physics" hold at extremely high temperatures or pressures. In somewhat more mundane fashion, many an investor has lost a bundle by attempting to predict future interest rates or stock trading prices on the basis of past or present trend.

Much effort is directed toward heat removal from semiconductor devices. A "natural" corollary of this seems to be that power-device behavior can be expected to improve with refrigerated cooling. Consider bipolar transistors: It is true that leakage currents continue to decrease with the lowering of junction temperature. Moreover, the thermal conductivity of silicon is about ten times better at liquid nitrogen temperature than at room temperature. As an extra bonus, the electrical conductivity of lead and bonding wires is much better at liquid nitrogen temperature than at room temperature. Unfortunately the efficiency of the emitter-base "charge gun" in a bipolar transistor diminishes with cooling so that the current gain at liquid nitrogen temperature is only a small fraction of that exhibited at room temperature. Thus, extrapolation falls down—it is not true that if some heat removal is beneficial, much more is necessarily better.

Enter the power MOSFET. Its temperature behavior is shown in Fig. 2-19. Its operation does not depend upon injection of minority charge carriers. Unlike the bipolar transistor it becomes a "new creature" at liquid nitrogen tempera-

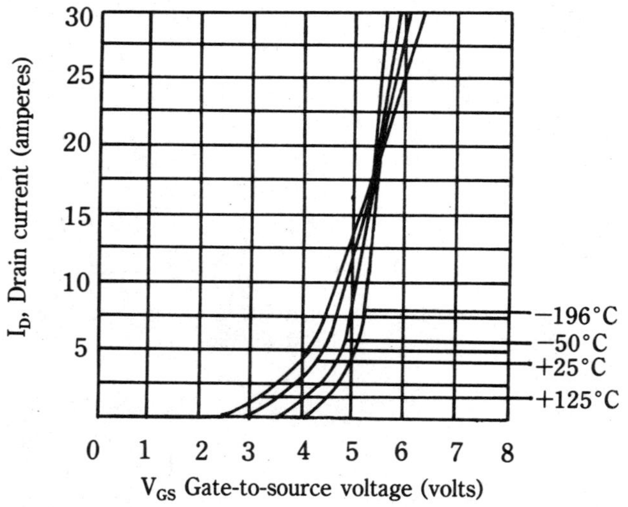

2-19 Transfer characteristics of a power MOSFET vs temperature.

ture, where its allowable power dissipation can be four or five times its rating at ordinary temperatures. Advances in the practical application of cryogenics have made it sometimes worthwhile to consider refrigerated operation. Indeed, passive components—capacitors, magnetic-core devices, and resistors—can, via appropriate design and selection, provide satisfactory performance at liquid nitrogen temperature and below. Although both Schottky and junction diodes display higher offset voltages, their other characteristics more nearly approach those of the ideal diode. These matters have particular relevancy to space communications. Interestingly, both extrapolation and experiment show that the power MOSFET can yield good performance all the way down to the temperature of liquid helium! A nice thing about depressed temperature operation of power MOSFETs is that the ON resistance, R_D, decreases along with temperature.

The "mysterious" demise of Schottky diodes

The Schottky rectifier has come into prominence as 5-volt, high-current power supplies have experienced increased demand by computer systems. Obviously, the offset voltage of a diode rectifier figures prominently in the dissipative losses of such supplies. For example, a single pn-junction diode might produce a forward voltage drop of 1.0 volt at, say 80 amperes. The resultant 80 watt loss is then a significant part of the 540 watts supplied to the load (5 volts times 80 amps). A single Schottky diode, used instead, could drop, typically about 0.65 volts, thus incurring a forward-drop loss of only 52 watts. The increased efficiency gained is obviously worthwhile. If you home in on other losses attending the use of the Schottky rectifier, such as reverse-leakage dissipation and dissipation in an often needed snubber network, the figures would have to be tempered somewhat. Also, at high frequencies, pn diodes can have significant switching losses from reverse-recovery action. Nonetheless, the cited example remains valid for comparison purposes.

The relatively low forward drop of the Schottky diode is a compelling factor in its selection, but the popular notion that reverse recovery is practically instantaneous often leads to trouble. To be sure, the Schottky diode does not store minority charges in the manner of a pn diode. Being a majority-carrier device, it should, indeed, be able to perfectly follow an alternating current wave. Unfortunately, it exhibits a storage phenomenon of a different kind. It displays a considerably higher self capacitance than does a pn diode of similar current rating—about five times as much. The net effect is that the Schottky diode can be likened to a fast-recovery pn diode, rather than an instantaneously responding device. (A switching transistor still "sees" a diode remaining ON when it should be OFF.) Moreover, the capacitance of the Schottky diode produces a side effect that often is responsible for "mysterious" destruction of the diode. Even worse, when the Schottky diode is used as a freewheeling diode, both it, and the switching transistor, might abruptly be relegated to the scrap heap.

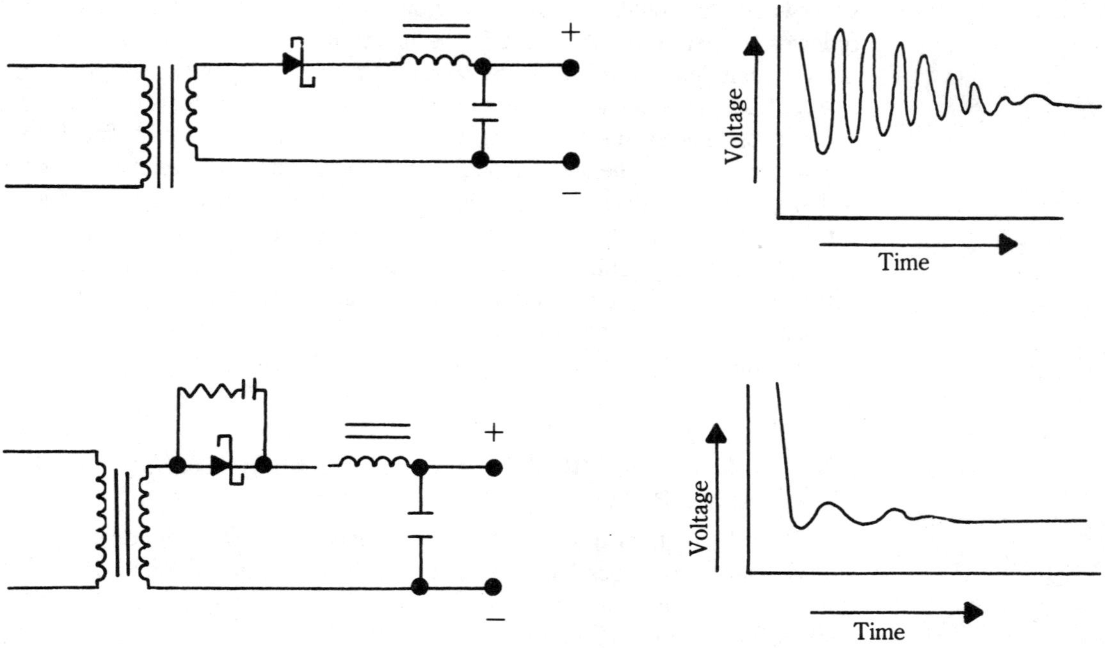

2-20 Mysterious demise of Schottky diodes—protective technique.

When used as a rectifier, the capacitance of the Schottky diode can resonate with the leakage inductance of the transformer feeding it. The voltage excursions of this shock-excited oscillation can destroy the Schottky diode. This is because Schottky diodes have low tolerance to reverse voltage. Figure 2-20 illustrates the situation and also shows the pacifying effect of a snubber-network connected across the diode.

3
What a meter is the wattmeter

ALTHOUGH ELECTRONIC TECHNOLOGY DERIVES ITS ROOT SUPPORT FROM basic principles of electrical engineering, it has often proven remarkable how proficient you can become with neglect of underlying theory, even if it had once been mastered. Of course, it is nice to have others do the actual engineering work for us and make practical implementation simple and easy via a complex IC with a few external terminals. The pros and cons of this ''tinker-toy'' approach will not be argued here. Suffice it to say there are situations where you must resort to classical concepts of electrical behavior to properly interpret circuit or system performance.

With regard to power supplies, remember the AC power input from the utility line is not generally the product of AC input voltage and AC input current as might be given by voltmeter and ammeter readings. This is because in practical power supplies, the line voltage and the line current are not in phase with one another. Even if the line voltage and line current were both sine waves, the in-phase condition would require the absence of reactance in both the power supply and in the load. With real world supplies and their loads, the product of line voltage and line current gives us the quantity called *apparent power* and then only if the voltmeter and ammeter respond to rms values. The improtance of this statement stems from the fact that the input or line current to a power supply is, more often than not, a nonsinusoidal wave.

The measurement of apparent power is useful providing you do not interpret it as true power, and providing you also have access to an instrument that will measure true power regardless of phase angle or waveshape. Having then made measurements of both apparent and true power, it follows from basic AC theory that the power-factor is equal to apparent power divided by true power. The closer to unity this is, the lower the excessive current required from the AC line by the power supply will be.

Although you are dealing with measurements unfortunately neglected by both vendor and user, often to their mutual dismay, it will be shown that there has been all along a remarkable instrument for measuring true power. Sadly, most of us have forgotten how to use it, and many have never known.

Why efficiency isn't always efficient

Some insight into the notion of efficiency of AC-operated equipment, such as a power supply, can be gleaned from the following example that brings out the importance of measurements involving true power, apparent power, power factor, as well as the significance of waveshape.

Consider two power-supply vendors, A and B. Vendor A advertises a one kilowatt supply with a claimed efficiency of 91%. Vendor A has determined that overall circuit losses in his supply are 100 watts. The operating efficiency was then calculated from

$$\frac{\text{Output}}{\text{Output} + \text{Losses}} = \frac{1000}{1000 + 100} = 0.91 \text{ or } 91\%$$

Vendor B also advertises a one kilowatt power supply, and also has determined the summation of all switching losses, conductive losses, and component losses to be 100 watts. Vendor B actually measured input power to be 1100 watts while the load consumed was one kilowatt. Then vendor B calculated operating efficiency from

$$\frac{\text{Input} - \text{Losses}}{\text{Input}} = \frac{1100 - 100}{1100} = 0.91 \text{ or } 91\%$$

Enter the utility company that says "a plague on both of your houses." This utility company claims that the 91% operating efficiency is misleading because of the low power factor of these supplies—about 0.60. (See Fig. 3-1.) The utility company contends that their AC feeder lines, transformer, fuses, and circuit breakers must cope with a supply current considerably higher than is indicated by a one kilowatt equipment with an operating efficiency of 91%. It is as if, the utility company states, the efficiency of the equipment was 90% × 0.60, or 54.6%. Even worse, the claim continues, the low power factor is partially the result of distorted line current. This generates harmonics that manifest their energy as EMI; the peaked current wave also plays havoc with circuit breaker behavior.

It is sometimes overlooked that the phase displacement of sine waves is not the only way a low power factor can result. The interaction between a sine wave and a nonsinusoidal wave produces a low power factor even though there is negligible inductance or capacitance in the circuit. Thus, phase-controlled SCR or triac supplies can produce notoriously low power factors. This is evident from the graph of Fig. 3-2.

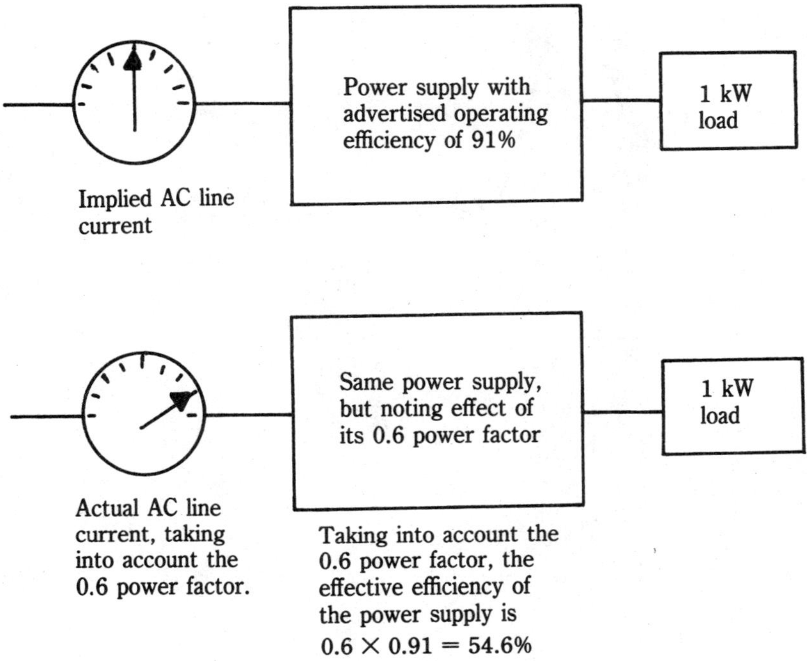

3-1 Efficiency is meaningless if power factor is neglected.

Rather than contend with either the vendors or the utility company, it is much better to make the requisite measurements of performance so that the true operating situation can be determined. To do this, a readily available analog wattmeter is needed, as is a little time to learn (or relearn) how to use it correctly.

The analog wattmeter

The measurement of AC input power and the determination of power factor is very important to anyone becoming involved with the performance of power supplies. Much misunderstanding has resulted from vendors claiming high efficiency from their equipment, but neglecting to tell the user that operation was also accompanied by a low power factor. As is well known, utility companies object to low power factor operation because higher current must be supplied than is the case with unity power factor. Moreover, low power factor operation of power supplies also evokes safety considerations, as well as increased EMI problems. Of course, if you are dealing with just a single small power supply, these factors might not assume much importance. However, with large supplies, or with a number of small ones, you should be able to measure true power, and apparent power; you should then be able to determine the power factor from these measurements.

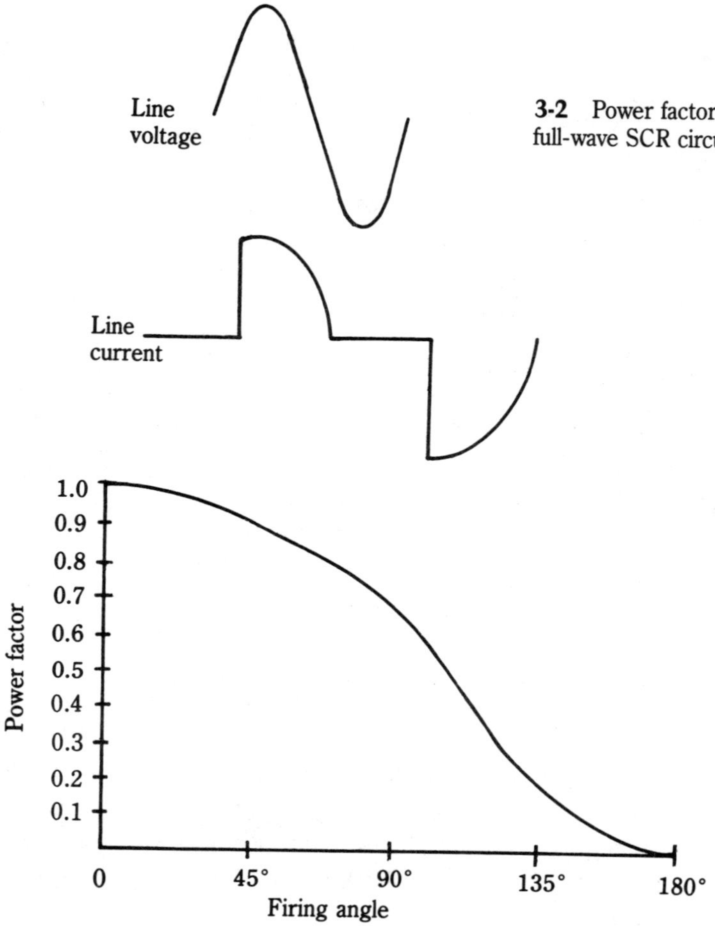

3-2 Power factor vs firing angle—triac or full-wave SCR circuit.

To measure apparent power, "old-fashioned" analog AC voltmeters and ammeters are satisfactory. Conversely, you can use rms responding digital meters. An elegant way to measure true power is to make use of a modern digital wattmeter. The trouble here is that these instruments are quite expensive. Despite their accuracy and convenience, they do not yet enjoy widespread usage, as do the digital voltmeters and ammeters.

Generally, it will be necessary to employ the analog wattmeter for true power measurements. Fortunately, these tend to be rugged and precise instruments and have long been used in 60 Hz power engineering. Unfortunately, little data is available regarding their use at higher frequencies. But the real obstacle is that electronics people are often unfamiliar with the proper use of this instrument. It happens that there is more to using the wattmeter than merely slapping it into the circuit or system.

This chapter will deal with the techniques and procedures that must be followed to not only obtain optimum accuracy, but to safeguard the instrument

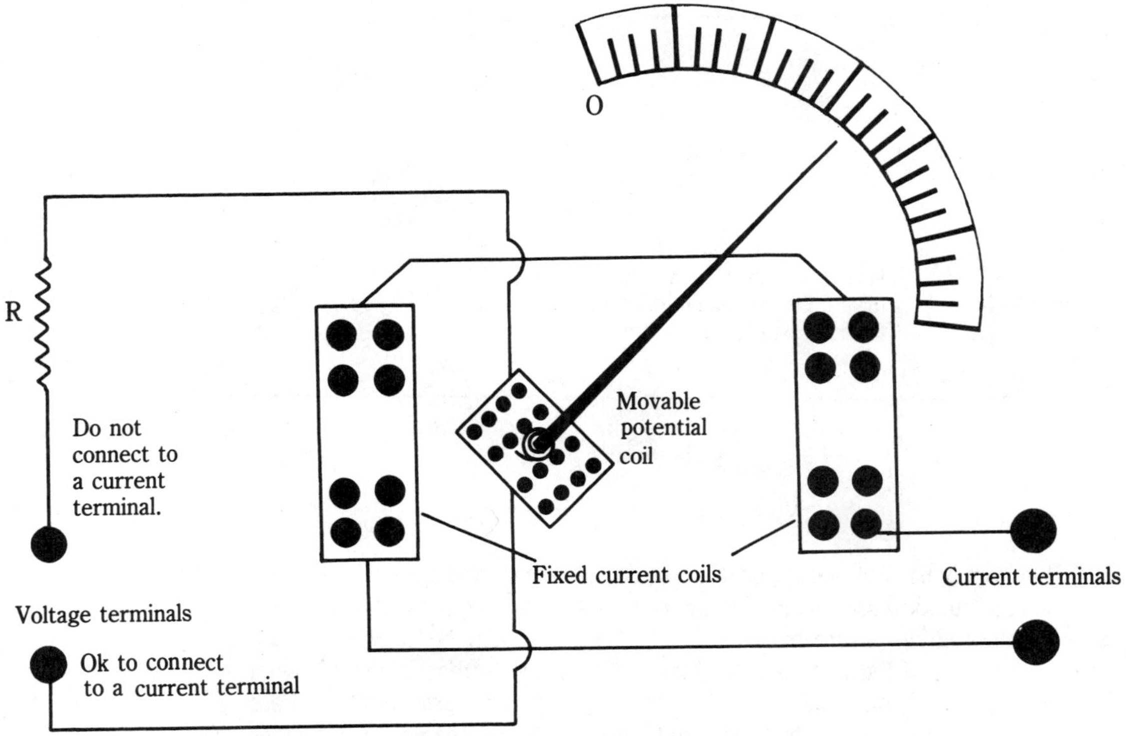

3-3 The electrodynamic wattmeter.

from damage. The interpretation of the readings in three-phase measurements will be discussed, as will the possibility of using the analog wattmeter at higher frequencies. Interestingly, it will be shown that this venerable instrument can be "just what the doctor ordered" for dealing with the nonsinusoidal currents consumed from the power line by modern electronic power supplies.

It will be seen that even if you gain the use of a sophisticated electronic wattmeter, much insight into the nature of power measurement will be obtained from an understanding of the principles involved in the use of the basic analog wattmeter.

The operation of the analog wattmeter is premised upon the electrodynamometer principle. Thus, the movement differs from the familiar D'Arsonval type used in DC voltmeters and ammeters. The major difference is that there is no permanent magnet as in the wattmeter, although there are still two interacting magnetic fields. But, in the wattmeter, both fields are supplied by the circuit undergoing measurement. The wattmeter movement is shown in Fig. 3-3. Inasmuch as both coils simultaneously experience polarity reversal in AC measurements, the net electromagnetic torque is always exerted in the same direction. In this respect, you are reminded of the similar situation prevailing in an AC universal motor, where continuous rotation in one direction takes place despite the reversals of the interacting magnetic fields.

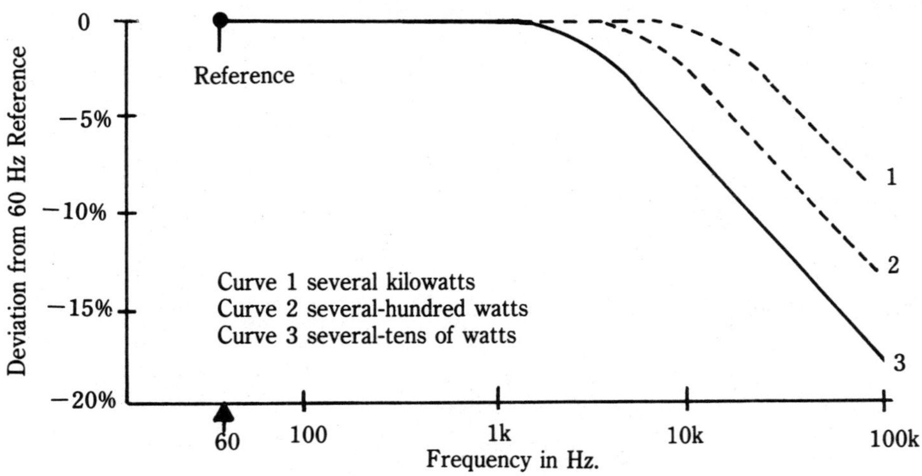

3-4 Wattmeter readings vs frequency.

It generally has not been appreciated that, notwithstanding their 60-Hz calibration and intended use, analog wattmeters have wideband response, enabling their application to electronic power supplies. This can be seen from the response curves of Fig. 3-4. These curves are composite responses of several wattmeters tested at various power levels. They depict general trends and are intended for guidance only. Note that extended response characterizes higher power levels. This is primarily due to the fact that both current and potential coils exert less inductive effect at higher powers. The current coil either has a shunt, or has fewer turns; the potential coil has higher series resistance.

Not shown on the graph is the fact that the response is substantially the same at DC as it is at 60 Hz. However, below a certain minimum frequency—on the order of a few Hz, the reading would no longer be steady, despite the steady reading at zero Hz (DC).

The nature of AC power

Figure 3-5 depicts current, voltage, and power in a simple resistive AC circuit. Interestingly, the power curve is double the frequency of the current and voltage sine waves. It, too, is a sine wave, but it is entirely above the axis of symmetry of the current and voltage curves. Perhaps you are not accustomed to thinking of power in this way. It has certain advantages, however. It will enable you to dispense with much mathematics pertinent to AC circuits. Graphic illustrations often lubricate the mental processes. There is nothing unrealistic or abstract about this double-frequency power curve—if you listen to the response of a photocell exposed to light from an incandescent lamp, there will be little doubt that the tone you hear is 120, not 60 Hz. A little consideration reveals why this is so. The instantaneous power resulting from the action of the current and voltage sine waves is, at any given time, their product. When both current

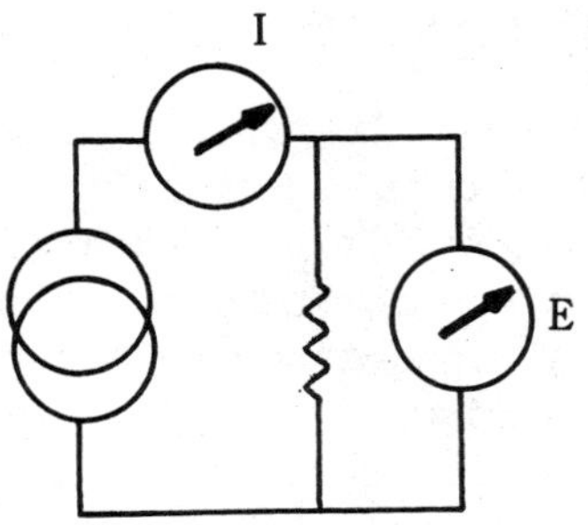

3-5 Power in a purely resistive AC circuit.

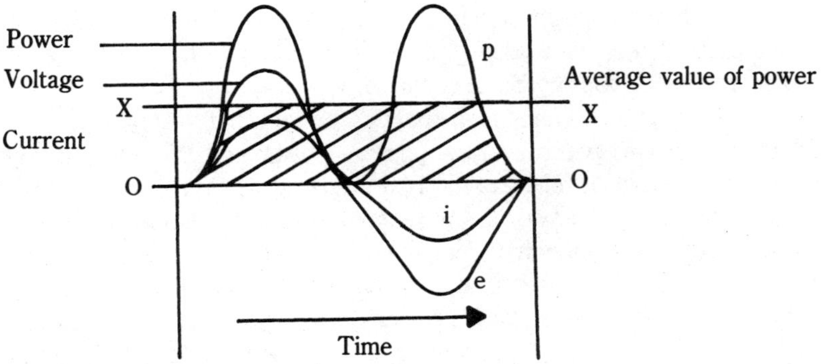

and voltage are positive, their product is positive. It is also true that their product is positive when current and voltage are both negative. That is, the product of two negative quantities is, itself, positive. You can see how the power curve doubles the frequency of the AC supply.

The best feature of the graphical format of Fig. 3-5 is that the power curve is what the wattmeter "sees." It is, indeed what any load or device experiences that is exposed to in-phase current and voltage such as exists in a resistive circuit. Except for unique situations, such as the photocell mentioned, you remain blissfully unaware of the nature of the power curve. This is because you generally are interested in the net effect of the power curve, or more specifically, its average value. It just happens that most power-consuming devices have a great deal of inertia and thereby average out the undulations of the power curve. Thus, you expect to feel a certain sustained heat from an electric heater, rather than 120 heat pulses per second. (The incandescent lamp is also a good example, for its light is quite steady—it takes a photocell or sensitive instrumentation to reveal the relatively small amount of 120 Hz variation.) Another way of saying all this is to state that the frequency response of most devices that convert AC electrical power to another form is not fast enough to follow the undulations of the power curve.

Without resorting to mathematical derivation, you can see in Fig. 3-5 that the average value of the power curve is its axis of symmetry, xx. This is the quantity you would like to have your wattmeter indicate. It makes sense, for the power areas above and below xx will just fill in the shaded area.

The concept of negative power

The situation illustrated in Fig. 3-6 is representative of the typical AC circuit—power is dissipated in a resistance, but the overall circuit is both resistive and reactive. In this particular case, an inductance is included to provide reactance. Capacitance would have served the same purpose, the essential difference then being that current would lead voltage in the waveform diagram rather than the converse situation.

The significant feature of the waveform diagram is that, unlike the situation in the purely resistive circuit, power is periodically negative. What, indeed, is the interpretation of such negative power? You can answer this from three vantage points. It has graphical, mathematical, and wattmeter implications. Although these three "viewpoints" all stem from common phenomena, insight is sharpened by considering them separately.

Graphically, it can be seen by inspection that the net average power will be the difference between the average positive power and the average negative power. This being so, it follows that a purely reactive circuit such as shown in Fig. 3-7 dissipates zero power in the reactive element—the positive and negative power waves being of equal area. The negative power is not a graphical illusion, nor is it a mathematical abstraction. Its "hardware" reality will be demonstrated shortly. The negative power represents energy previously stored for one-half cycle in the reactive element, then returned during the subsequent (negative) half cycle to the AC source. Thus neither a pure inductance nor a pure capacitance consume net power over any number of complete cycles.

Mathematically, the same concept is vested in the equation $P = EI (\cos \Theta)$. That is, the power dissipated or consumed in an AC circuit is the product of rms voltage and rms current and the cosine of the angle between applied voltage and resultant current. (The rms values of voltage and current are those indicated by ordinary meters.) Cosine Θ is known as the power factor. The mathe-

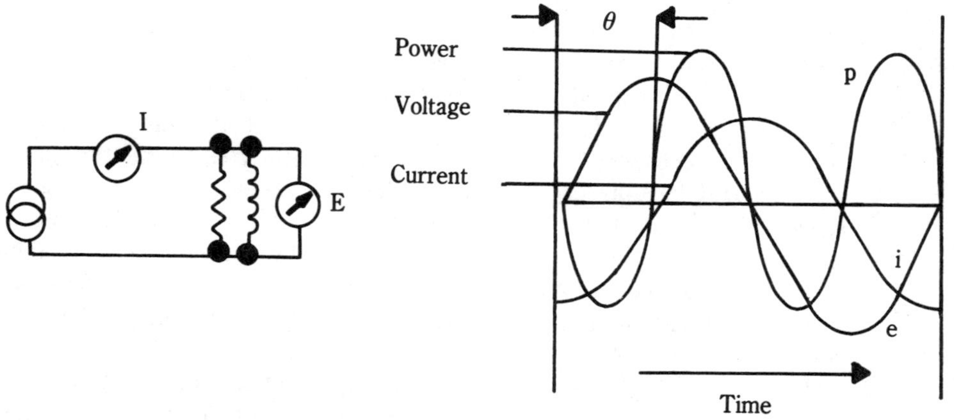

3-6 Power in a simple reactive AC circuit.

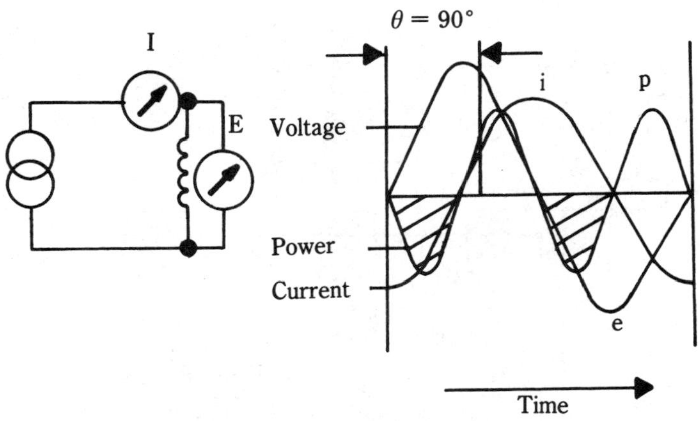

3-7 Power in a purely reactive AC circuit.

matical derivation of this equation will not be given here. Suffice it to say that it is premised upon the interesting fact that the power wave is always sinusoidal. This is true whether the circuit is resistive, reactive, or as is most generally the case, partly resistive and partly reactive. If you have some means of knowing Θ, such as from oscillographic inspection, or from calculation when resistance and reactance are known values, $P = EI(\cos \Theta)$ conveniently gives true power in an AC circuit.

Practical insight into negative power

The "reality" of negative power can be demonstrated by an experiment such as is shown in Fig. 3-8. Here, the wattmeter is connected to a low frequency reactive circuit, on the order of several hertz. The meter movement will not have sufficient inertia to average out the variations in the power wave. Therefore, the pointer will be seen to more or less follow instantaneous power values. In particular, it will be seen that the pointer at times occupies the scale region between zero and the left pin stop or scale extremity. The meter obviously experiences negative torque at these times and you can correctly infer that power is being transferred from the inductance to the AC source, rather than vice versa. This not only demonstrates the concept and reality of negative power, but also the importance of mechanical inertia when the meter is used at frequencies it was designed for.

Interestingly, negative power can be indicated at ordinary frequencies where the inertia of the meter movement is operative. This can result from inadvertently transposing either the voltage or the current terminals. Such a backward indication does not necessarily reveal negative power in the circuit, but only that you have connected the meter incorrectly. However, when wattmeters are used in three-phase power measurements, one of the meters can actually produce a reverse indication even though it is correctly connected! To conveniently take care of such a situation, some wattmeters have reversing

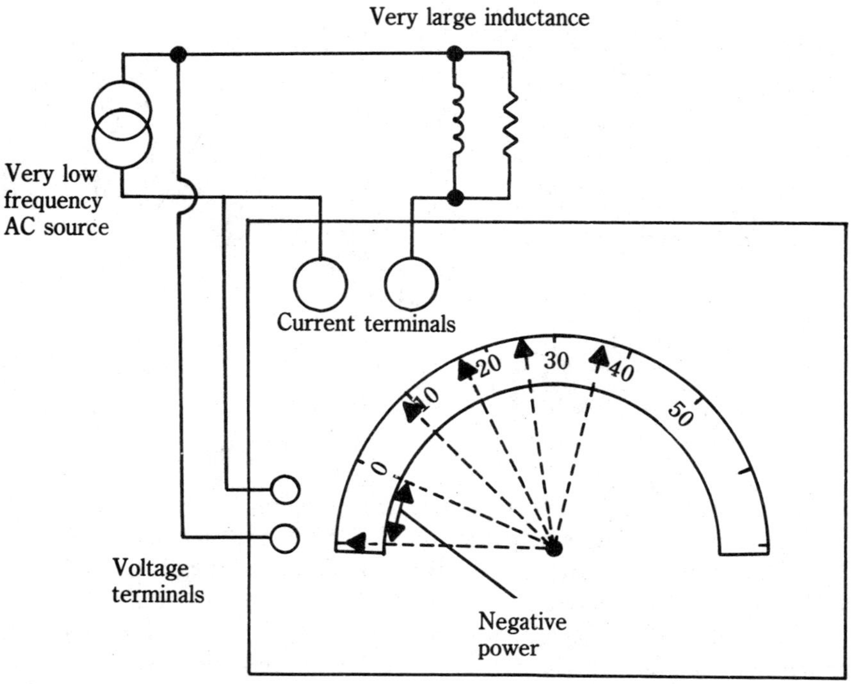

3-8 A "hardware" demonstration of negative power.

switches associated with their voltage coils. Although this enables an up-scale reading to be obtained, the observer must be careful to record the indicated power as a negative quantity when combining it with the indications of the other wattmeters. Although this situation is not the same as that pertaining to your low-frequency experiment, it reinforces the face that negative torque can be developed in the meter, i.e., the meter "knows" the direction of power flow.

Using the wattmeter with nonsinusoidal current

The electrodynamometer-wattmeter, when properly connected in an AC circuit involving sine waves of voltage and current, will indicate the true power dissipated or consumed in the resistive part of the load. In providing such indication, the wattmeter averages all instantaneous values of both positive and negative power. As you have seen, the presence of reactance—either inductive or capacitive—causes negative power for a portion of the time. The meter movement is in essence an analog computer that gives a readout of the equation $P = E \times I \times \cos \Theta$. As previously shown, Θ is the angular displacement between applied voltage and resultant current. Despite its venerable vintage, the wattmeter surely qualifies as a nifty instrument. The fact that it automatically takes into account voltage, current, and phase angle would endow it with a respectable IQ, were it animate rather than hardware!

It is appropriate here to mention another capability of this type of wattmeter. In accordance with the classical literature on the subject, the use of this instrument has been emphasized for sine wave circuits. In this day and age, however, your need to measure AC power can involve such nonsinusoidal current waves as are encountered in SCR, triac, and switching power-supply circuits. What then? Surprisingly, the electrodynamometer-wattmeter is often capable of providing reasonably accurate measurements in such circuits. The criteria is simply, can the meter measure sine waves up to three times the frequency of the nonsinusoid? If so, a useably accurate power reading can be obtained. If the meter is capable of accurate response from sine waves of five or seven times the frequency of the nonsinusoidal wave, so much the better! This reasoning is based upon the Fourier theorem that states that nonsinusoidal waves actually are composed of a sine wave fundamental frequency plus a number of harmonically related sine waves.

Where the wattmeter meets this stipulated requisite, you need not be concerned over the relative amplitudes or phase angles of the various harmonics. Again, the meter preempts the "smart" instrumentation of our modern era—it automatically indicates the overall power contribution from the fundamental and the harmonics. In practice this means that 60-Hz wattmeters are useful in measuring power in SCR and triac circuits.

Connecting the current and potential coils

The instrument you have investigated has, indeed, an "intelligence" of its own. You might suppose it is only necessary to slap the wattmeter into the circuit undergoing measurement and note the reading. Unfortunately, such procedure might, at best, produce an unacceptably inaccurate indication; at worst, it can damage or destroy the instrument. See how this comes about.

From the previous discussion, the current coil is supposed to monitor the load current. And the potential coil is supposed to sample the voltage across the load. In actual practice, only one, but not both of these requisites can be met. Depending upon how the current and potential coils are connected in the load circuit, either the sampled voltage will be too low by the IR drop in the current coil, or the current passing through the current coil will be too high by the current consumption of the potential coil. Specifically, the connection format of Fig. 3-9 enables the meter to "see" correct load current, but a higher than actual load voltage. Conversely, the coil connections used in Fig. 3-10 enable the meter to "see" correct load voltage, but higher than actual load current. In most instances, one connection scheme is best and the error might be negligible. But how do you know which connections to select?

The general idea is to employ the connections shown in Fig. 3-10 when load power results from high current and low voltage. Conversely, the connec-

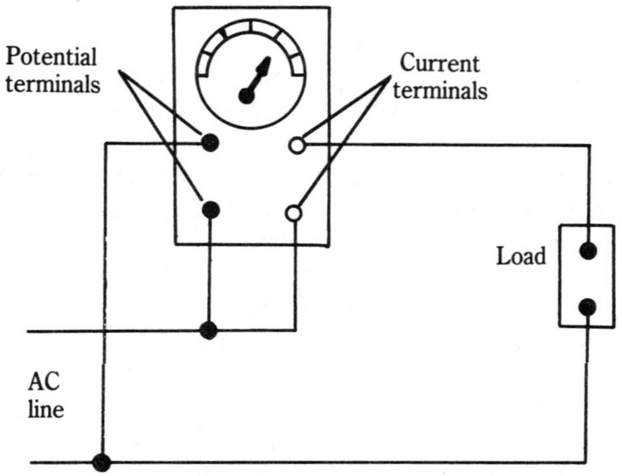

3-9 One of two correct connection formats.

tions shown in Fig. 3-9 should be used when the load power is the result of low current and high voltage. What actually constitutes "high" and "low" current and voltage levels depends on the meter and, in any case, cannot be stated easily in quantitative terms. If the resistances of the current and potential coils are known, as well as an approximate value of load current and load voltage, computations can be made to indicate the best connection format. Usually, intuition serves this purpose. Such intuition can be premised on such a rule of thumb that loads of less than one ampere accompanied by voltages greater than, say fifty volts, likely would be measured most accurately by connecting the wattmeter coils as shown in Fig. 3-9. Loads involving much higher currents and significantly lower voltages would, conceivably, be best measured via the coil connection format depicted in Fig. 3-10.

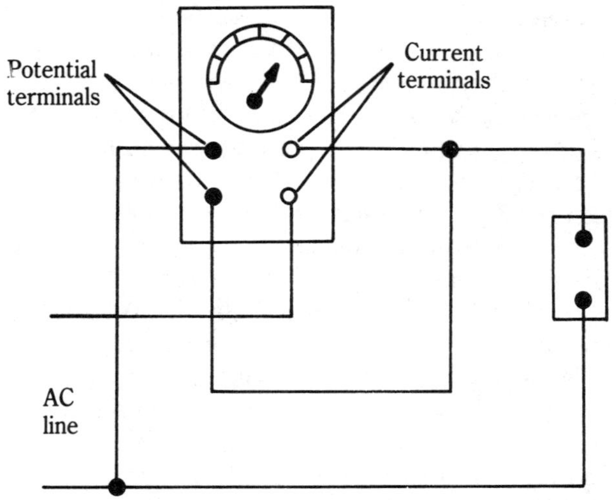

3-10 The second of two correct connection formats.

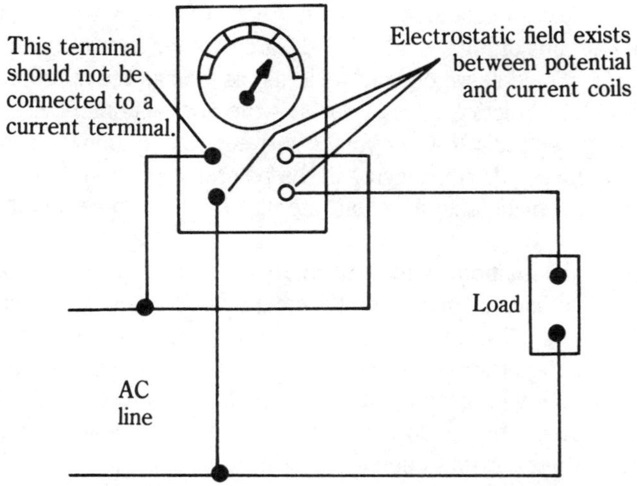

3-11 One of two incorrect connection formats.

cantly lower voltages would, conceivably, be best measured via the coil connection format depicted in Fig. 3-10.

More ado about the wattmeter connections

There are two additional coil connections that you might use when connecting the wattmeter in the load circuit. These are shown in Figs. 3-11 and 3-12. These can lead to even greater errors than merely selecting the wrong format

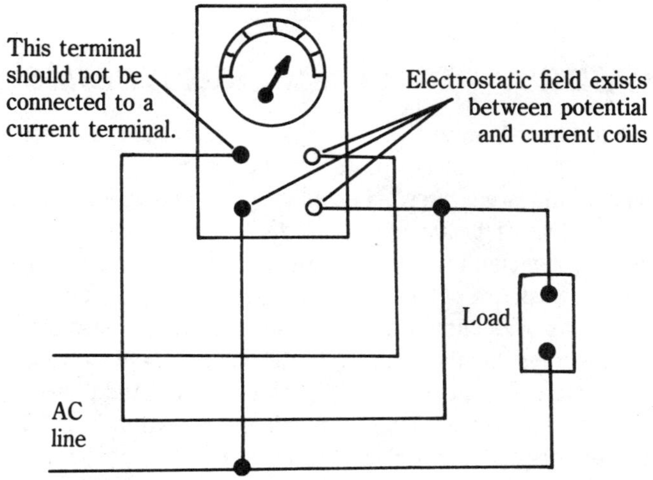

3-12 The second of two incorrect connection formats.

of the two connection schemes shown in Figs. 3-9 and 3-10. It might appear that the connections of Figs. 3-11 and 3-12 are counterparts of those in Figs. 3-9 and 3-10 and should serve equally well, just as long as the appropriate substitution is made. In practice, this is not so—the connections of Figs. 3-11 and 3-12 tend to develop *electrostatic forces* between the stationary and moveable coils that introduces additional error in the pointer indication. This is not obvious from the schematic diagrams, but stems from the physical construction of the meter.

To avoid this situation, some wattmeters only have three terminals. With such meters, you cannot inadvertently connect the terminals to produce this electrostatic effect. However, by the same token, you no longer have the option to select one or the other connection formats shown in Figs. 3-9 and 3-10.

Surely, by now you are ready to connect the wattmeter to the load circuit and measure power—or are you? Despite the precautions already cited, there remains yet another consideration if you wish to protect the meter. If too much current passes through an ammeter, or if too much voltage is applied to a voltmeter, these instruments will peg the pointer stop, giving visual evidence of the overloaded condition. Unfortunately, this is not true of the wattmeter. A given number of watts can derive from high voltage and low current, or from low voltage and high current. Thus, even though the pointer only deflects to, say midscale, either the current or voltage coil can be greatly overloaded. This can burn out the meter or produce insulation breakdown. Here, again, you must have a "feel" for the magnitudes of the currents and voltages to be measured and especially how they relate to the ratings of the meter.

It is clear that the wattmeter, though a "brainy" instrument, also has its areas of stupidity. It is obviously up to you to supply intelligence where it is missing. When this is done, the wattmeter becomes an exceedingly useful measurement instrument, one that can yield reasonably accurate readings despite power factor, frequency, or wave shape.

Three-phase power measurements

A discussion of the wattmeter lacks completeness without at least brief reference to its use in three-phase power circuits. This is particularly relevant because many of the larger supplies in industry are energized from three-phase AC lines. If the load is balanced, as is often the case, the single wattmeter method can be conveniently used, as shown in Fig. 3-13. The resistances, R, are equal to the resistance of the wattmeter's potential coil. (This can be predetermined by means of an ohmmeter measurement.) Regardless of power factor, the power consumed by the load (generally either wye or delta-connected transformer primaries) is given by multiplying the indicated power by three.

If two wattmeters are available, the scheme shown in Fig. 3-14 is useful, providing the meters are identical. If they are laid out as shown, there need be no confusion whether to compute power as the sum or the difference of the two readings. The rule is simple: If both meters produce upscale readings, the total

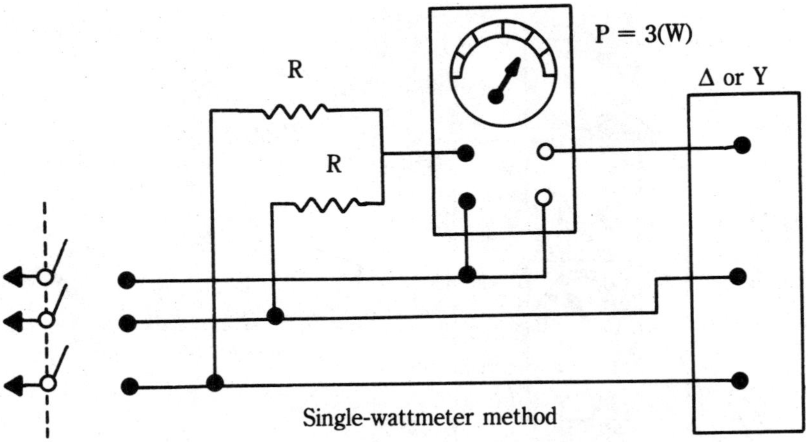

3-13 Single wattmeter method for measuring three-phase power.

power is the sum of these readings. If, however, one meter deflects backward, reverse its current-coil connections, then subtract the now upscale reading from that of the other meter. (If a balanced load is operating under unity power factor, both meters will produce the same upscale reading; if the power factor is 0.5, one meter will read zero, whereupon total power is read from the other meter alone.)

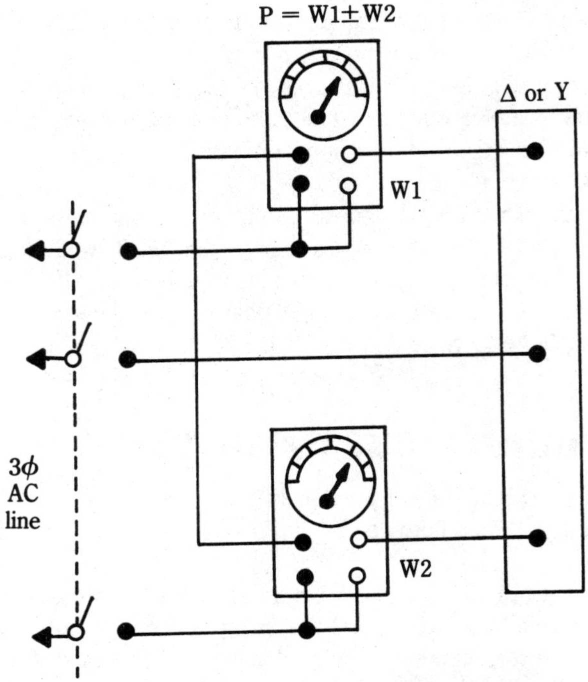

3-14 Two-wattmeter method for measuring three-phase power.

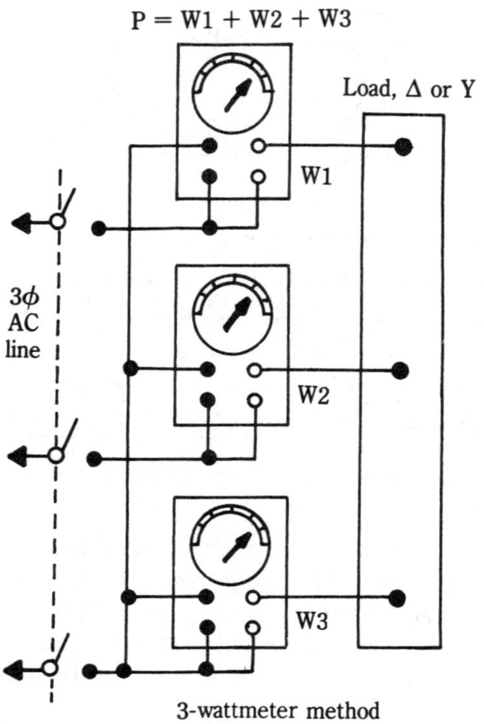

3-15 Three-wattmeter method for measuring three-phase power.

Three identical wattmeters can be used in a manner that is independent of power factor and that does not require a balanced load. Such a measurement circuit is shown in Fig. 3-15. In this case, the total power is always the simple arithmetic sum of the readings. Actually, the scheme remains valid even if the meters are not identical. Under such conditions, however, you will not observe identical readings for a balanced load.

Note that the situation of balanced load automatically infers the presence of a balanced line, for if line voltages are unbalanced, what would otherwise be a balanced load then operates in an unbalanced fashion too. It is interesting to observe that all of the described three-phase measurement methods work equally well for either wye or delta connected loads.

Notes on previous discussions

The following notes relate to the previously discussed topics and could well have been included there had it not been my intent to present basic principles, and to keep things simple.

With regard to Figs. 3-9 and 3-10 and the relevant discussion on measurement errors, it should be mentioned that many wattmeters are intended to be connected into the load circuit as depicted in Fig. 3-10 no matter what the relative current and voltage levels are. Such meters display no error from the current in the potential coil because they have a *compensating winding*. The compensating winding is wound with fine wire between the turns of the current coil and is connected in series with the potential coil. The ampere turns of the compensating winding subtract from those of the current coil by just the right amount to cancel the error that otherwise would be produced by the potential coil current flowing in the current coil. These are often three-terminal instruments and are very handy to use.

The single-wattmeter method of measuring three-phase power can be modified from the circuitry shown in Fig. 3-13. Specifically, it is feasible to dispense with the two resistances "R" when reasonably balanced line and load exist. Simply connect the potential coil to one of the phases sharing the current coil. Overall power will then be 1.73 times the meter reading.

An interesting and very useful feature of the two-wattmeter method shown in Fig. 3-14 is that the *power factor* of the load can be calculated from the power readings. To do this, use the equation,

$$\tan \Theta = (1.73) \frac{W1 - W2}{W1 + W2}$$

After tan Θ is obtained in this manner, a table of trigonometric functions is consulted and the corresponding value of cos Θ is recorded. Voila! Cos Θ is the power factor. Just make sure that due heed is paid to the algebraic signs of the two power readings. If the current terminals of a meter had to be interchanged, or if the "reverse" switch on such a meter had to be actuated to obtain an upscale reading, that reading is considered negative.

Finally, to prevent lost time and energy in reinventing the wheel, a *polyphase wattmeter* already exists! It makes use of the two-wattmeter principle, but the two potential coils are mounted on a common shaft. The torques thereby produced automatically combine power levels "W1" and "W2" so that the pointer of the meter indicates the overall power in the three-phase load. The connections for such a meter are shown in Fig. 3-16.

Surely, the dynamometer-type wattmeter was a pretty smart instrument in its day and you see that its IQ hasn't diminished with time!

The electronic wattmeter

The electronic wattmeter has the potential of ultimately displacing the classical analog wattmeter. Sometimes referred to as a "digital wattmeter," these more modern instruments can provide features of accuracy and measurement convenience not readily attainable from the analog instruments. A noteworthy feature of the electronic meter is that voltage, current, and power can be measured by the same instrument without altering the connections. When so used, power factor can be derived easily by simple computation from the relationship

$$\text{Power factor} = \frac{\text{True power}}{\text{Apparent power}}$$

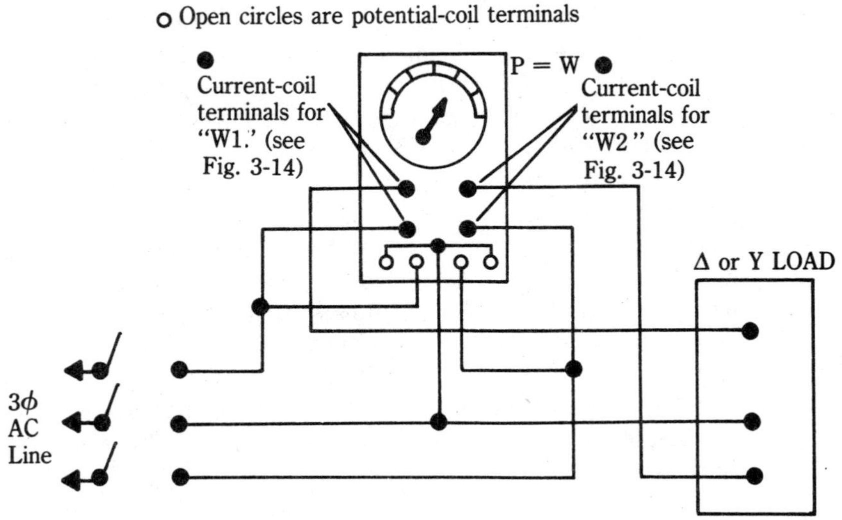

3-16 Measurement of three-phase power with a polyphase wattmeter.

or

$$\text{P.F.} = \frac{P}{E \times I}$$

Figure 3-17 shows one of such instruments made by the Clarke-Hess Communication Research Corp. In line with what has been stated, this instrument is called a digital volt-ampere-wattmeter, or more simply a V-A-W meter. Various models are available for instrumentation of broadband frequency ranges from DC to one MHz. By making use of external shunts or current transformers, currents from 50 microamps to 500 amperes can be accommodated. Full-scale voltages normally can cover the range of 100 millivolts to 1000 volts.

These instruments, unlike their analog predecessors, impart negligible loading on the circuits being measured. Thus, there is no need to take into account the voltage drop involved in current sensing, nor the current consumption from voltage sensing. This implies one standard way of connecting the instrument into the circuit being measured without regard to relative magnitudes of voltage or current. Also, unlike many of the hand-held "true" power meters, appropriate models of these instruments can handle DC components and DC levels. Thus, the true power reading is the overall rms power in all frequency components present in the circuit being measured.

Many options are available, such as IEEE-488 bus operation where both measurement and display modes can be remotely or computer controlled. With the progressive increase of operating frequency of switch-mode power supplies, the demand and popularity of such electronic meters are certain to grow. It is intriguing to contemplate that both the traditional and newer instruments are responsive to E × I × cos Θ (true power); the former meter does this via mechanical inertia and electromagnetic interaction, whereas the later meter

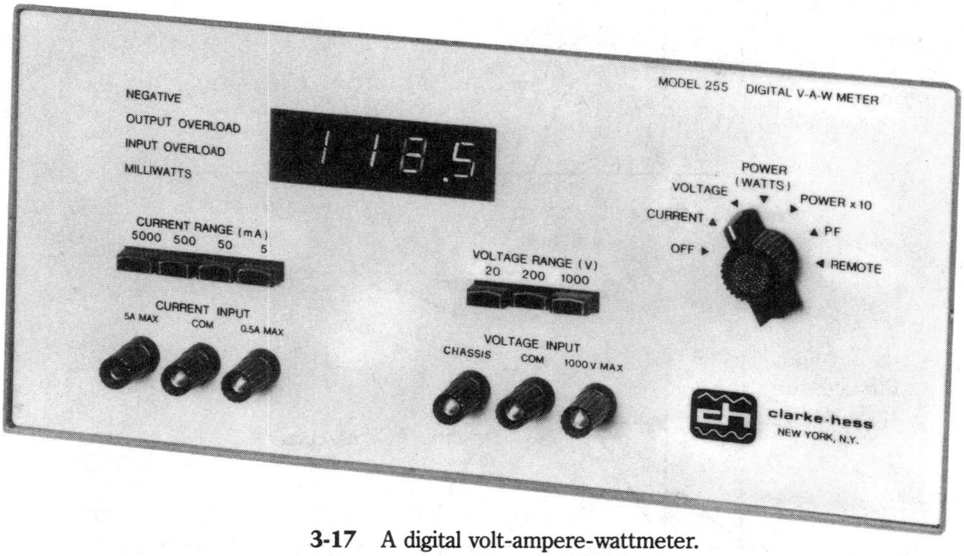

3-17 A digital volt-ampere-wattmeter.

accomplishes the same result by means of electronic sampling and processing. The principal obstacle to more widespread deployment of such instruments has been cost. Undoubtedly, this will be largely overcome in time.

A common pitfall with old or new instrumentation

In previous discussions dealing with the nature of AC power, it was shown that the power factor is the cosine of the angle between voltage and current. Thus, the power factor of a power supply would normally be construed to be the cosine of the angular displacement between input voltage and input current. This stems from the classical definition of power factor generally found in texts dealing with 60-Hz power technology. Such texts do not neglect to state, but often neglect to emphasize, that such a definition requires that both the voltage and the current be sine waves. In power supplies, this is not likely to be true. A not uncommon measurement mistake made in power supply evaluation is the application of the relationship $PF = E \times I \times \cos \Theta$ when the AC line current, I, is not sinusoidal. The ease in which voltage and current waves can be observed on the oscilloscope screen renders such an erroneous computation almost natural. Keep in mind, however, that the angle seen on the scope between nonsinusoidal waves is not Θ. See Fig. 3-18.

The fact that an SCR or a triac working into a purely resistive load can exhibit a low power factor should make you wary of applying $PF = E \times I \times \cos \Theta$ to nonsinusoidal waves. Yet, the concept of assigning a power factor value to equipment operating with nonsinusoidal current or voltage is a very useful one. This might appear to pose a conflict and requires a bit of thinking.

It turns out that a more universal definition of power factor is that it is the ratio of true power to apparent power. Thus, the power factor of a power supply ordinarily would be determined from,

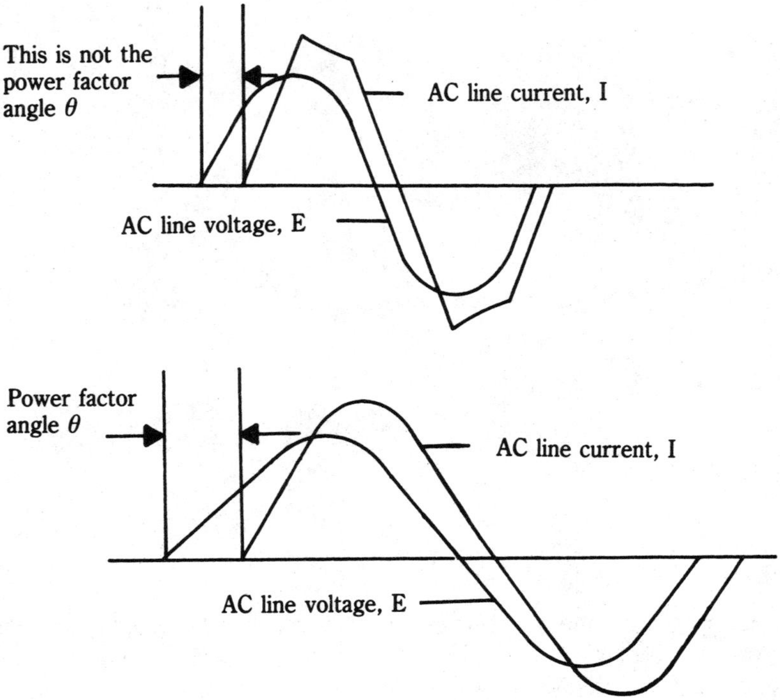

3-18 Dealing with equivalent power factor of nonsinusoidal waves.

$$PF = \frac{\text{True input power}}{\text{Input voltage} \times \text{Input current}}$$

As has previously been stated, True input power must be measured by either an analog or an electronic wattmeter. The essence of such meters is that they are "blind" to waveshape and phase. Input voltage and input current must be measured by rms-responding instruments so that, again, waveshape does not enter into the reading. This equation for power factor is useful for all waveshapes of current and voltage, even though engineering texts frown on its nonsinusoidal application.

Now deal with the implication of "power factor" as applied to power supplies and other equipment where both voltage and current may not be sine waves. A power supply determined to have a certain power factor from the ratio of true to apparent power behaves as if input voltage and input current were sine waves with an angular displacement of Θ degrees. Or, looking at the situation in reverse, such an "as if" power supply would operate at a power factor equal to $\cos \Theta$. This makes sense to the utility company.

The nice thing about the electronic wattmeter is the convenience with which power factor can be determined. This is because power, voltage, or current can be read by merely setting panel controls. Indeed, power factor, itself, can be read directly in this manner.

Miscellaneous considerations

Most electrical equipment, including power supplies, that does not operate at unity power factor tends to exhibit a lagging power factor where line current lags line voltage by the angle Θ. A leading power factor is also possible in which line current leads line voltage by the angle Θ. Considered by itself, a leading power factor is just as bad as a lagging one. However in the real world where most lines are subject to lagging power factors, the presence of equipment operating with a leading power factor can be a welcome situation. This stems from cancelling effects of leading and lagging reactive volt amperes. Thus if two equipments operating with equal, but opposite power factors consumed the same number of reactive volt amperes and were connected to an AC line, that line would "see" a net unity power factor.

The electronic wattmeter is "user friendly," intellectual analysis needed is minimal. For example, leading or lagging power factor, or negative power are indicated by LEDs. Moreover, you need not have any concern over "right" or "wrong" terminal connections—both, voltage drop across the current terminals and current loading by the voltage terminals are negligible. Wideband shunts and current transformers, as well as voltage dividers are available to extend measurement ranges by simple powers of ten. Options peculiar to electronic instrumentation also are available. These include parallel four digit BCD output, remote-control provision, IEEE-488 talker/listener board, and an isolated analog output where one volt corresponds to a full scale display reading.

A possible concern when first using an electronic wattmeter is the fact that it must be plugged into an appropriate AC supply to derive its operating power. However, more than adequate safety margins are incorporated into the design of such an instrument as the Clarke-Hess Model 255, as well as other models, such that there is no interaction with the system undergoing measurement. Voltage isolation is always safely higher than the highest voltage involved in any proper measurement procedure.

On the other hand, the need for AC operating power could be construed as an inconvenience in some instances—those where the accuracy and sophistications of an electronic instrument are not deemed necessary. It would seem that an option allowing operation from an internal rechargeable battery could be worthwhile. At this writing, such a provision has yet to appear on the market.

The rf wattmeter

Up to now, books dealing with power supplies have not been appropriate sources for information about radio frequency techniques. Because of the ongoing trend toward higher switching rates, future authors will be compelled to merge energy conversion and rf practices to provide meaningful insights into power supply operation. The several hundred kHz to one MHz region already has seen considerable commercial exploitation; it appears likely that switchers operating in the several MHz range offer desirable features, and it is not clear how high a frequency borders the area of diminishing returns. In any event, high frequency supplies already partake many of the aspects of rf equipment.

The measurement of rf power is desirable for the sake of circuit optimization. Generally the input to a high-frequency power supply will be taken from the 60-Hz utility line, and the output will be the desired DC. In between, however, the high-frequency switching circuitry will have to be designed, evaluated, and serviced. It is here that one accustomed only to low-frequency components, construction, and instrumentation can go astray. In particular, the measurement of power can involve difficulties. The venerable power-line wattmeter is not useable at rf, and various models of electronic wattmeters might or might not be.

There is another commonly used analog meter that is useful in measuring rf power. This meter shows maximum load power when its terminations are 50 ohms resistive. This might seem ridiculous from the vantage point of low-frequency practice, where the wattmeter can be inserted into a circuit without regard to impedance level. Those versed in the art of manipulating rf systems deal with this feature by using suitable transformers or impedance-transforming networks. These are so implemented that the rf wattmeter "sees" input and output terminations that are as near as possible to being 50 ohms resistive.

The rf wattmeter also serves as a "power factor" meter, except that standing-wave ratio (SWR) displaces the concept of power factor in rf practice. The arithmetic is different, to be sure. However, like power factor in low-frequency circuits, standing-wave ratio can denote the effect of reactance; as with power factor, maximum transfer of power occurs when generator and load "see" matched resistances, and net reactance is zero. Unity power factor and unity SWR both denote this desirable condition. However, power factor is zero for a purely reactive circuit, whereas SWR is infinity.

The basic reason why an rf wattmeter can't be just a scaled-up version of a low-frequency meter is that space, length, and distance become important parameters at rf. This means that a wire carrying current will not necessarily have the same current at different distances along it. Similarly, the voltage with respect to ground can vary greatly as you progress along the length of the current-carrying wire. Indeed, SWR implies the ratio of voltages measured at peaks and valleys along such a wire. The higher the frequency (the shorter the corresponding wavelength) the closer together these peaks and valleys will be.

The current-carrying wire alluded to actually represents a transmission line with the return conductor being the ground plane. In actual practice, rf is usually conveyed via deliberate transmission lines such as coaxial cable or twisted pair of wires. All transmission lines exhibit a so-called characteristic impedance. For practical purposes, when a transmission line is terminated by a generator internal resistance and by a load resistance both equal to its characteristic impedance, there will be maximum transfer of power to the load. Moreover, under such a matched condition, the current and voltage in the transmission line will not change with distance; that is, the SWR will be one. Inasmuch as coaxial cable with a characteristic impedance of approximately 50 ohms is much used in

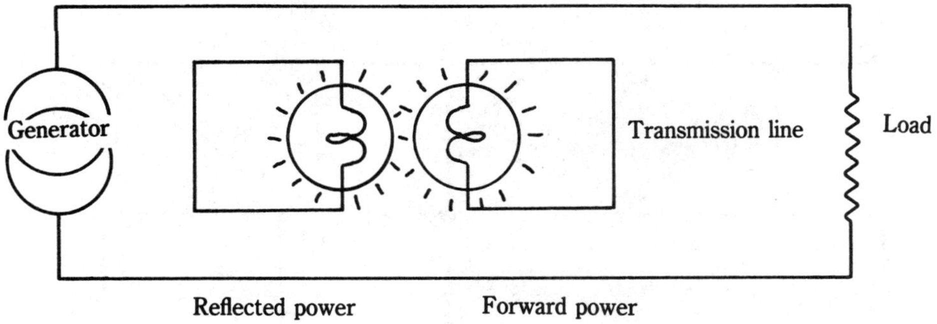

3-19 Setup demonstrating basic principle of the rf wattmeter.

rf practice, the rf wattmeter is commonly designed to accommodate this impedance level.

It goes without saying that the ideal 50-ohm terminations generally will not be encountered in practice. This means that the SWR will be greater than unity. Such a higher than one SWR has another implication—it corresponds to the fact that power has both a forward and a reflected component. That is, part of the power heading for the load is reflected back to the generator. Fortunately, the rf wattmeter can measure forward power, and at the flick of a switch, reflected power. This being the case, the actual power absorbed by the load is forward power minus reflected power. You'll find in a mismatched system, a high value of forward power might be indicated, but the reflected power is likewise high. In an ideally matched system, the forward power reading is the actual power absorbed in the load because the reflected power reading will be zero.

With these matters in mind, it is easy to demonstrate the basic principle of the rf wattmeter; it involves the ability to measure the forward and the reflected components of power in a transmission line interposed between generator and load. Depicted in Fig. 3-19 is a simple rf power system made up of a generator or source of rf power, a transmission line, and a load. Additionally, two pick-up loops are oriented as shown. In this simple arrangement, reflected and forward power are proportional to the intensities registered by the lamps. This is not the only way to directionally sample power in the transmission line, but it is essentially the basis for various modifications.

A "bare bones" version of a commercial rf wattmeter is shown in Fig. 3-20. Here the inner conductor of a section of 50-ohm coaxial cable is coupled to a toroidal pick-up inductor. The rectified rf is applied to an analog meter, which is often a 100 microampere DC meter. Not shown in this simplified circuit, there commonly would be several switch-selectable attenuators to enable accommodation of different power ranges. Such attenuators could be simply shunts or multiplier resistances for varying the sensitivity of the meter. A more formal name of such an instrument is the Breune in-line SWR/wattmeter.

It is evident that the voltage standing wave ratio, or SWR, is not directly indicated by the meter, but rather forward and reflected power. However, SWR is readily calculated as follows:

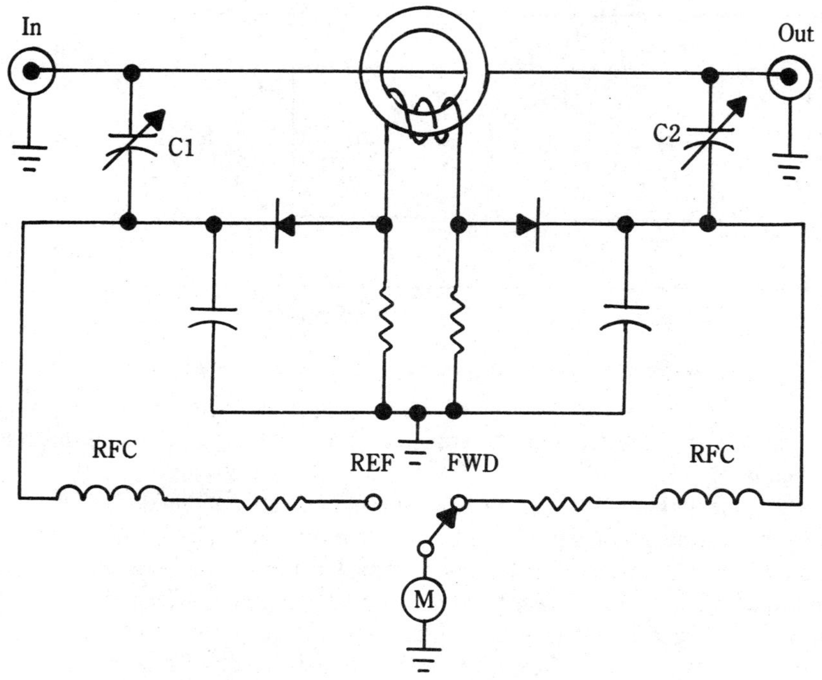

3-20 A "bare bones" version of a commercial rf wattmeter.

First, an intermediate parameter, the reflection coefficient, is determined from

$$k = \sqrt{\frac{Pr}{Pf}}$$

where: Pr is the indicated reflected power
Pf is the indicated forward power

then,

$$SWR = \frac{1 + k}{1 - k}$$

It is well to point out that SWR, is conventionally used, represents VSWR.

A useful accompaniment of the rf wattmeter is a 50 ohm resistor designed to be nearly purely resistive at rf and which can, at least for a reasonable time, safely dissipate a rated power level. This resistor is commonly known as a "dummy load" and can incorporate various means of heat removal.

An example of a way in which the rf wattmeter might be used to help optimize the performance of a power supply involving rf for energy transformation is as follows: Suppose it is required to convert a 48 volt DC source to 1 volt DC at a moderate power level, say 100 watts. The emphasis is on compactness and

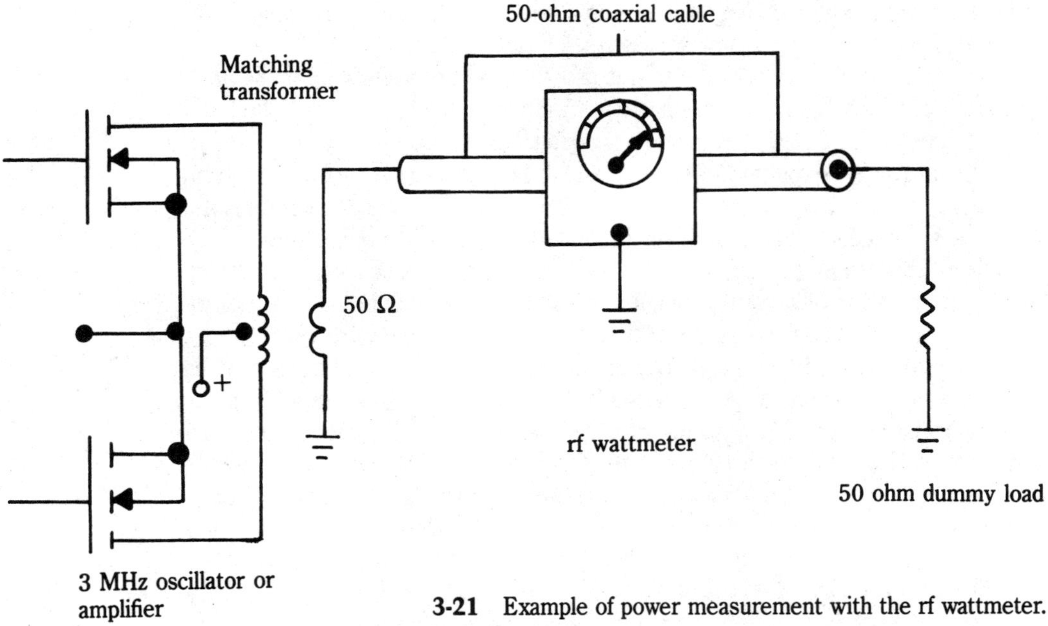

3-21 Example of power measurement with the rf wattmeter.

on operating efficiency. It is decided to explore the possibility of using a pair of power MOSFETs in a push-pull self-excited oscillator operating in the 3 MHz region. If such a scheme can provide efficient rf power, then it would appear feasible to accomplish low loss rectification with a Schottky diode.

Inasmuch as a number of interdependent factors enter the operation of such an arrangement, it is generally wise to employ both theory and empirical methods. An experimental breadboard version of such a supply would enable optimization of excitation, bias, resonating techniques, transformer design, waveshape, device selection, and other factors not always amenable to mathematical calculation. Thus, in a setup such as shown in Fig. 3-21 you could find the best combination of operating conditions to achieve maximum output. The nice thing about such an approach is that it provides the actual output power that can be derived from the oscillator (or amplifier). Indeed, such a procedure is much easier to do than either write about or read about. At rf, experimental changes in the transformer are not at all difficult to make. Most likely, you would not be interested in SWR, but only in the net power delivered to the load.

Electronically controlled load for test convenience

The use of a wattmeter presupposes that the data recorded will be correlated to certain load conditions of the power supply undergoing tests. Typically, you might wish to gather information so that AC power input, efficiency, and power factor can be determined for, say 20, 50, 80, 100, and 120% of full load. Often, the requisite load resistances are not readily obtainable, and makeshift loads can be characterized by approximation or instability. Measurement convenience is best served by an adjustable load with sufficient thermal inertia so that a reasonable time can be allotted to observation and recording of instrument readings. Of course, the longer such a test load can endure a given power dissipation the better.

A popular laboratory solution to this problem has been a test jig made up of a number of paralleled bipolar transistors with adjustable base drive. The basic idea in building such an "electronic rheostat" is to provide as much heat-removal hardware as practically feasible, and to hope for equal-current partitioning among the paralleled power transistors. The latter operating condition is not always easy to achieve, although considerable help is forthcoming from the use of emitter-ballast resistors. Even so, bipolar transistors are susceptible to thermal runaway; if one of several paralleled transistors begins to hog a disappropriate share of total load current, all of the transistors can be destroyed.

The situation is much more favorable when power MOSFETs are used as load absorbers. Unless grossly overloaded, power MOSFETs are immune to thermal runaway. Indeed, by virtue of their positive coefficient of output resistance, they tend to self-correct any tendency for current hogging. Although best paralleling results from matched units, the emphasis on uniformity is not as severe as it is with bipolar transistors. Also, a small battery suffices for gate-control voltage regardless of the number of units.

The basic circuit for adjustable load is shown in Fig. 3-22. Because of the bridge rectifier, this simple and versatile test jig can be used to load inverters as well as converters and DC power supplies. At higher frequencies, the bridge elements should be Schottky diodes rather than pn diodes. In using this test jig,

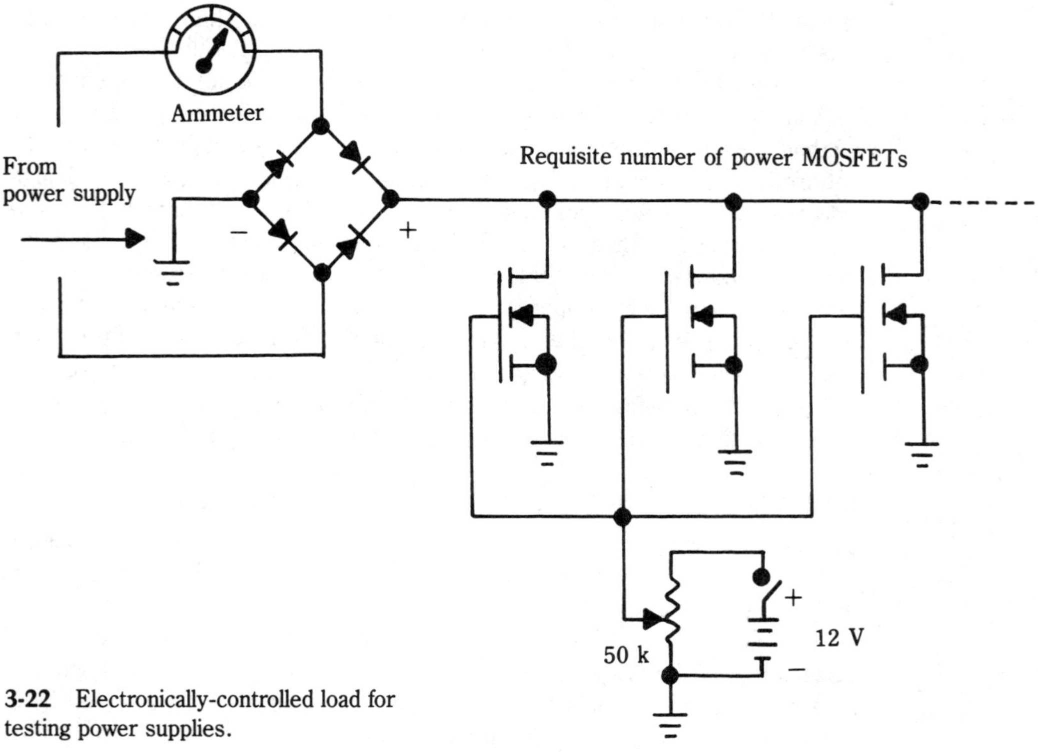

3-22 Electronically-controlled load for testing power supplies.

do not confuse the "ground" symbols with the system ground of the power supply undergoing test.

Power and power-factor information from waveshapes

A commonplace situation with power supplies is that the impressed voltage from the AC line is sinusoidal, whereas the consumed current is not. Useful estimates of operating conditions can be made under such circumstances by means of geometric relationships. For commonly encountered waveforms, these relationships often can be found in tables. Also, simple calculations can be made from observation of waveform features. Inasmuch as the applied voltage is a sine wave, the objective is to obtain the rms value of the current wave. Then enough information is available for calculation of power and power factor.

Observation of the AC current waveform is accomplished with the oscilloscope in conjunction with some kind of current-sampling device. This can take the form of a very small resistance inserted in the AC line, a current transformer, a Hall-effect element, or most conveniently, by special clamp-on current probes made by the oscilloscope manufacturer.

Consider the situation depicted in Fig. 3-23. The applied AC voltage is sinusoidal; the resultant current consists of narrow rectangular pulses symmetrically located with respect to the voltage peaks. Despite this "in-phase" relationship between the voltage and current, the power factor must be less than unity because the harmonics contained in the current pulses are not interacting with similar harmonics from the voltage wave. It turns out that the power factor is nearly equal to twice the square root of D, where D = t/T. Thus, for such narrow current-pulses, Power Factor = $2\sqrt{D}$. The accuracy of this relationship is primarily dependent upon the constancy of the voltage over the width of

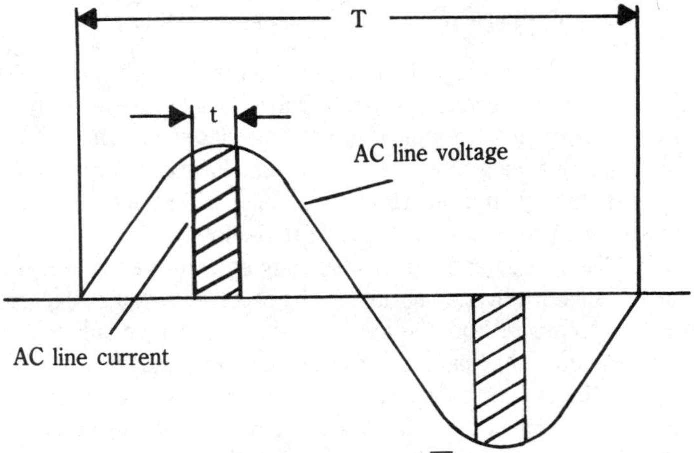

$$\text{Power factor} = 2\sqrt{d} = 2\sqrt{\frac{t}{T}} \text{ where } D = \frac{t}{T}$$

3-23 Power factor calculation from sine-wave voltage and narrow current pulse.

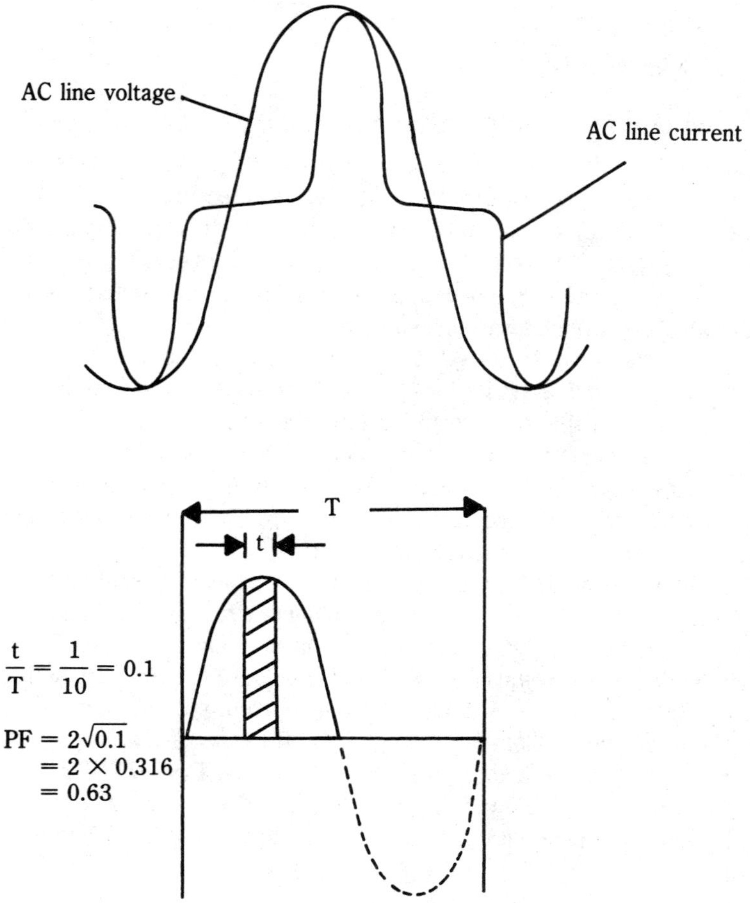

3-24 Practical example of power-factor determination for a power supply.

the current pulse. Therefore, this method is most applicable to relatively narrow current pulses. However, the current pulse does not necessarily have to be rectangular; an approximately-equivalent rectangular shape can be derived from the actual shape by seeing to it that both shapes have the same area.

To show that this is not merely a mathematical abstraction from the classroom, apply it to a practical situation. Figure 3-24a represents an oscilloscopic display of AC line voltage and current typically encountered in power supplies. The distorted current wave is because of the rectifier and filter capacitor input. There is some departure from the idealized situation shown in Fig. 3-23. Nonetheless, some useful "ball parking" can be accomplished.

In Fig. 3-24b, the current wave has been redrawn as a rectangle with the same area as the original wave. Additionally, symmetry has been restored by shifting the rectangle slightly to the left. This is permissible because of the essentially constant voltage level of the sine wave in the vicinity of its peak. It turns out that $D = t/T = 0.1$. Application of the formula for power factor of narrow current pulses yields the value, 0.63. This is interesting, for this is, indeed, the power factor of typical power supplies using a rectifier and a single filter capacitor at the input. It must be emphasized, however, this waveshape relationship is only applicable to narrow current pulses situated at or near to the peak of the voltage sine wave.

Additional useful operating information pertaining to waveshapes is listed in Table 3-1. Here, you deal with line current waveshapes encountered in three popular control circuits for power supplies. These circuits are the triac, the back-to-back SCR configuration, and the single SCR associated with a rectifying bridge. These circuits are used extensively in phase-controlled power supplies for electric motors. There is also considerable use as preregulators for power supplies associated with electronic loads. When so used, they substitute for a heavy and costly constant-voltage (ferroresonant) transformer. Unfortunately, there has not always been awareness of the low power factors that accompany this type of control at low conduction angles.

It should be appreciated that the data in this table pertains to a resistive load. The presence of reactance in the load alters the current waveshapes, so some judgement might be required for certain operating conditions. Many loads can be essentially resistive, especially at 60 Hz. The unity power factor operation in the load for *all* conduction angles is the result of the identical waveshape of load current and load voltage. Keep in mind that electronic system loads involving rectifiers and filter capacitors will exhibit the peaked-current waveform previously discussed. Under such conditions, the data in Table 3-1 might be useful, but will not be accurate.

In addition to the phase-control waveforms of Table 3-1, other frequently found power-supply waveforms are shown in Table 3-2, together with their pertinent parameters. Particularly relevant to power and power factor determinations are the rms values of these waves. It is assumed the waves are observed and measured with an oscilloscope. The relationships are straightforward enough, but there are a couple of possible pitfalls: Note that the rms value of the half-wave rectified sine wave is not one-half that of the full-wave rectified sine wave. This is an instance where the ''knee-jerk'' reaction of common sense leads you astray.

A second possible pitfall involves the rectified square wave; it is not true that its average, peak, and rms value are equal. Such a statement applies to the true square wave with equal half cycles about a zero horizontal axis. The reason this error can be made by even those who know better is that the oscilloscopic display does not always clearly reveal the difference between the two waveforms. A surefire way to distinguish between the two is to operate the scope in its DC, rather than its AC measurement mode. Otherwise the coupling capacitor(s) in the vertical-amplifier circuit will make the rectified square wave resemble a true square wave.

The rectangular-pulse relationships are useful for many switch-mode supplies employing pulse-width modulation. From the maximum duty cycle to be encountered in such a supply, the rms value of the maximum current through

Table 3-1. Power and power-factor relationships for three control circuits.

Line voltage waveform (sine waves)	Current conduction-angle	Line and load current waveform (load voltage waveform)	Line current (rms value)	Apparent power	True power	Power factor at AC line	Power factor at load
Em	180°	Im	0.707 Im	0.500 Emim	0.500 Emim	1.000	1.000
	150°		0.697 Im	0.493 Emim	0.486 Emim	0.985	1.000
	120°		0.634 Im	0.448 Emim	0.402 Emim	0.897	1.000
	90°		0.500 Im	0.354 Emim	0.250 Emim	0.707	1.000
	60°		0.313 Im	0.221 Emim	0.098 Emim	0.443	1.000
	30°		0.120 Im	0.085 Emim	0.014 Emim	0.170	1.000

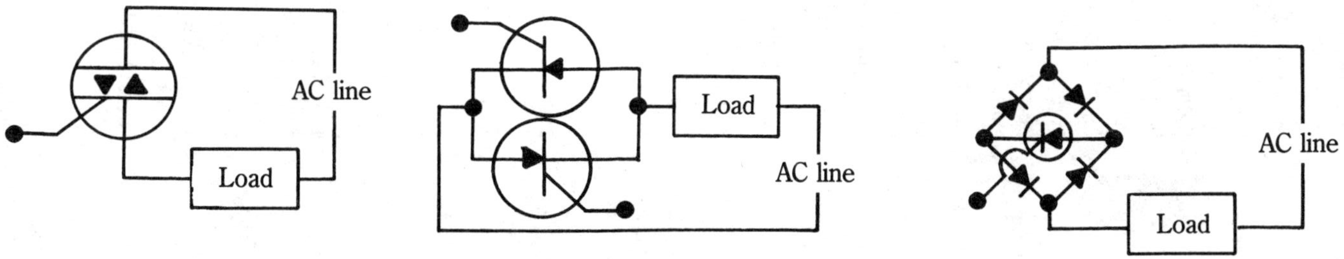

(A) (B) (C)

Table 3-2. Useful parameters of power-supply waveforms

Waveform			Multiply pk value py
Sine:	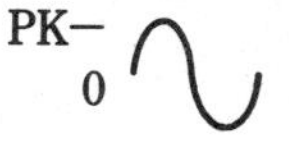	pk-pk	2.000
		0-pk	1.000
		rms	0.707
		avg	0.637
Rectified sine: (full wave)		pk-pk	1.000
		0 – pk	1.000
		rms	0.707
		avg	0.637
Rectified sine: (half wave)		pk-pk	1.000
		0-pk	1.000
		rms	0.500
		avg	0.318
True square:	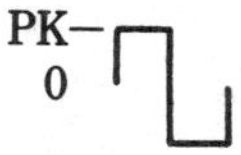	pk-pk	2.000
		0-pk	1.000
		rms	1.000
		avg	1.000
Rectified square: (half wave)		pk-pk	1.000
		0-pk	1.000
		rms	0.707
		avg	0.500
Rectangular pulse:	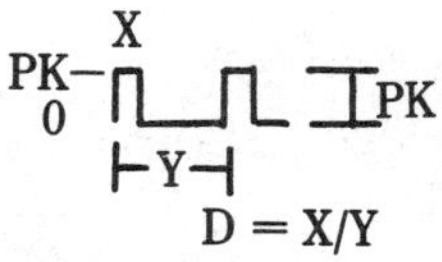	pk-pk	1.000
		0-pk	1.000
		rms	$D^{1/2}$
		avg	D
Triangle:	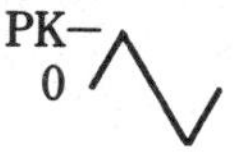	pk-pk	2.000
		0-pk	1.000
		rms	0.577
		avg	0.500

the power switch can be determined. From this and the manufacturer's specifications on R_D, or on voltage drop, the all-important power dissipation in the power switch can be closely estimated. The value thereby derived will often have to be tempered by an estimation of switching losses and, in bipolar transistors, by additional losses in the base-emitter junction.

Not shown in Table 3-2 is the full-wave rectified square wave. This is because such a ''waveform'' would simply be a constant DC level at value PK. Thus, ideal full-wave rectification of a true square wave would require no filtering! Practically, you must deal with trapezoidal, or near-square waves, and must cope with nonlinearities, storage phenomena, and other nonideal behavior of rectifying diodes. Nonetheless, less filtering is needed for the full-wave rectification of a reasonably good square wave than for most other nonsinusoidal waveforms. The PWM waveform of switching power supplies can impose particularly severe demands on the filter system.

It is easy to see that the oscilloscope is a useful adjunct to the wattmeter in the determination of power and power factor in power supply systems and circuits.

4

Interesting power supply items

The PolySwitch®—a self-healing plastic fuse

It seems only natural to think of plastic substances as electrical insulators. From our knowledge of most such subtances, what little conductivity exists tends to get better with increasing temperature. It might, accordingly, come as a surprise that such natural behavior has been completely reversed in a polymer plastic material developed by the Raychem Corp. This company markets a device it terms a PolySwitch® that conducts exceedingly well at normal and nominal temperatures, but that abruptly becomes a near insulator at a certain elevated temperature.

Such a device can be said to have a positive temperature coefficient, just the inverse of the common negative-coefficient device, the thermistor. The PolySwitch is intended to be used as a protective device for electrical equipment and has potential application in power supplies, inverters, and converters. See Fig. 4-1. It can be likened to a self-healing fuse, because when the fault condition responsible for opening it is remedied, it is ready again to conduct normal current levels. It is also superior to circuit breakers in that its action is faster and its restoration is automatic.

A nice feature of the PolySwitch is that it does not recycle unless the fault condition that caused it to open is removed. Thus, the equipment isn't forced to periodically sample the overload. Unlike a fuse or circuit breaker, the PolySwitch passes a small leakage, or holding current when in the ''open'' condition. The self-healing effect of this holding current keeps the device latched in its high-resistance state. Reversion to high conductivity requires removal of the circuit fault condition, and a cooling-down time of twenty seconds or so.

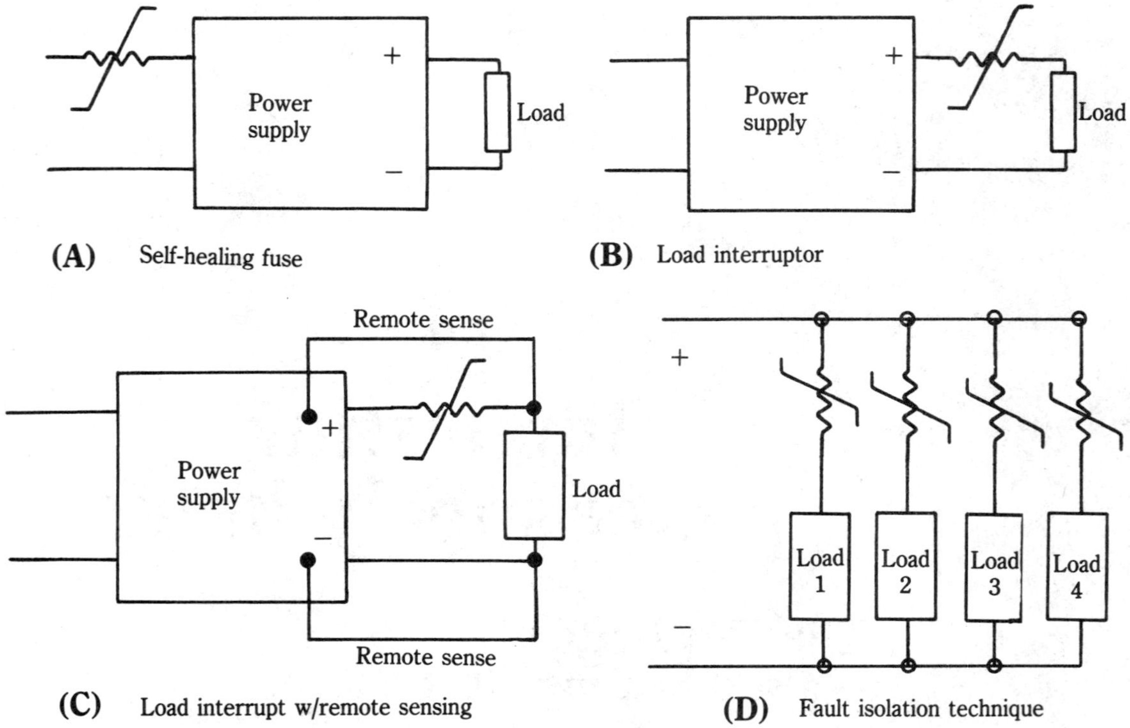

4-1 Applications of the polyswitch® circuit protector.

The PolySwitch resembles in appearance a ceramic disk capacitor. Conductivity can be almost ''metallic''—on the order of tens of milliohms. For many practical situations the degradation of power-supply regulation will be negligible, especially with the use of remote sensing, assuming the PolySwitch is placed in an output lead. The PolySwitch also can be placed in an input lead to the power supply in the same manner as a line fuse. The important thing is to select the appropriate PolySwitch in terms of current and voltage ratings. As a general rule, the trip current can be expected to be on the order of 150% of the rated hold current. Most PolySwitch circuit protectors are recognized by the Underwriters Laboratories.

Planar magnetic components

Various techniques have been alluded to that enable extension of switching rate. These techniques have included circuit techniques, such as resonant-mode operation; better power switches, such as power MOSFETs and upgraded bipolar transistors; and constructional methods—those generally used in high-frequency rf work. It also has been shown that the passive components must be given due consideration. Diodes must be of the fast-recovery type, and often preferably Schottky types, which have no minority carrier storage problems.

Capacitors have severe restraints imposed on them in terms of self-resonances, ESR, and ESL, as well as the usual high-frequency losses from dielectric hysteresis. But, not much has been said about inductors and transformers.

This is largely because more or less acceptable performance has been forthcoming by adhering to the core and winding methods generally used in rf practice. Ferrite cores can be optimized for minimal losses, and can be reliably specified for required flux densities. To a considerable extent, this also applies to powdered iron and permaloy materials. Then, too, toroidal and cup core geometries have provided both fabrication and operational advantages. Copper losses have been minimized by using Litz wire, multiconductor cable, hollow tubing, or foils to circumvent skin-effect losses.

Further optimization of performance from magnetics components can be attained via a fresh approach. Instead of merely placing windings on available core shapes, it is necessary to start from scratch and do what is dictated by high-frequency considerations. This implies a laminated structure of spiraled windings and surrounding ferrite. The flat ''sandwich'' produced excels conventional structures at high frequencies. Leakage inductance and copper losses in the flat inductor or transformer are very low. Thermal coupling to a heat sink, if needed, is inordinately good (convection cooling usually suffices). Reproducibility and repeatability are precise. Packaging and mounting characteristics are easy; space and volume are conserved. The use of such components in production is cost-effective, inasmuch as no winding need be done.

Multisource Technology Corp. has been a pioneer in the development of planar magnetic components specifically intended for use in high-frequency power supplies. The use of such pretooled magnetics helps make high-frequency power supply design less of a black art, and is recommended for both the experimenter amd the production line.

The optimum way to apply postregulation

The salient feature of all switch-mode supplies—both pulse-width modulation and resonant-mode types—is operating efficiency. This is reflected in compact, lightweight equipment involving minimal thermal problems. The overall effect of these factors on cost is invariably favorable, becoming more so for supplies with higher power capability.

Notwithstanding dramatic improvements in switch-mode technology, it has remained an open secret that nothing beats a linear regulator where truly clean output is needed for the load. This, at once, suggests the use of postregulation where a linear regulator is interposed between the switcher and the load. Such a technique should, in principle, give the user the best of both worlds—reasonably high efficiency and smooth operating current. A nice thing about the concept is that the linear regulator operates from a constant voltage source, greatly relaxing its electrical and thermal design problems.

Indeed, many switch-mode supplies have incorporated postregulation. There have been some problems. In an extremely competitive market, it is not always easy to add too many "luxury" items. Then, too, there has been the problem of degradation of overall efficiency. This is because commonly available three-terminal linear regulators tend to have too high a drop-out voltage or too high a quiescent current, or more generally, both. What has been needed is three-terminal linear regulators dedicated just for such postregulation applications.

Linear Technology Corp has pioneered such linear regulators, bearing the designations LT1083, LT1084, and LT1085. These regulators handle different current ranges—from 3 to 7.5 amperes. Their drop-out voltages are relatively low—as low as 0.9 volt, and their quiescent current can be as low as 50 microamps. Drop-out voltage is the difference between input and output voltage and is very significant in a postregulation application because the power dissipation of the linear regulator increases as the square of its drop-out voltage. Drop-out voltage as a function of temperature is depicted for the LT1083 in Fig. 4-2. Quiescent current is similarly important because it represents current never delivered to the load. Here, too, you are confronted with the square relationship, i.e., dissipation of power in the linear regulator goes up as the square of quiescent current.

In light of this situation, it behooves the designer aiming for postregulation of a basic switch-mode power supply to seek out dedicated three-terminal linear regulators that have been optimized to dissipate minimal power in such an application.

High-voltage protection for linear power supply

Despite the dramatic progress that has been made in switch-mode power supplies, the linear regulator has by no means been relegated to obsolescence. It remains superior to switching supplies in the domains of quiet operation (electrically speaking) and in tight regulation. Linear regulators often are more cost-

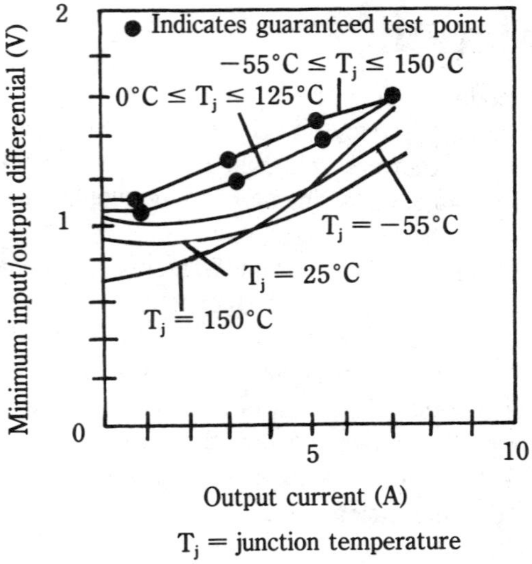

4-2 Dropout voltage of the LT1083 adjustable linear regulator.

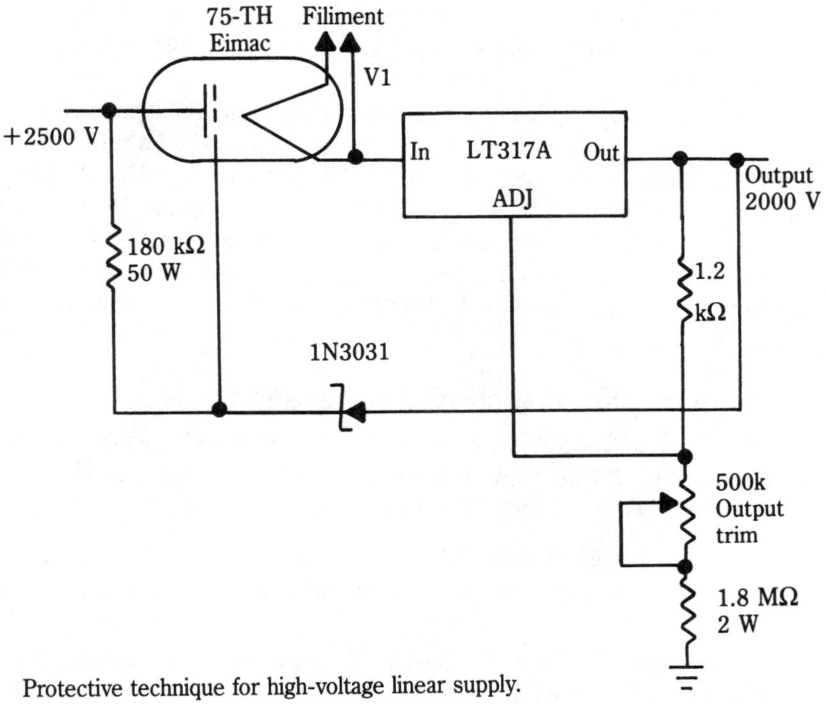

4-3 Protective technique for high-voltage linear supply.

effective than equally applicable switchers, especially for low power levels, say below 20 watts. Also, there tends to be a convergence of convenience, simplicity, and economy when linear techniques are used for high-voltage regulators. A difficulty, however, has been the matter of short-circuit protection.

The arrangement shown in Fig. 4-3 supplies a regulated 2000 volts to a 600-watt load via an inexpensive three-terminal linear regulator. At the same time, it is short-circuit proof. At first thought you might suppose that this circuit is just another version of the tube semiconductor circuits in which the tube functions as the series-pass element. In this scheme, however, the tube serves in a different role—as a ready and willing ''absorber'' of both short-circuit current and high voltage. Because of its electrically rugged characteristics, and its forgiving nature, the tube can perform under conditions that destroy semiconductors.

The arrangement of Fig. 4-3 is feasible because of the nongrounded ADJ terminal of the LT317A regulator. Under normal service, the grid of the tube is sufficiently positive to permit passage of full-load current. The regulation of the power supply is not assisted by the presence of the tube because its grid is not within the feedback loop. Also, the Zener diode is normally off. All is well because of the safely low voltage maintained across the input–output terminals of the LT317A regulator.

Without the tube, a short-circuited output would instantly destroy the regulator. Because of the tube, such a short circuit turns on the Zener diode, thereby applying about +30 volts to the grid. Although this increases the conductivity through the tube and semmingly adds fuel to the fire, the current path through the grid-filament circuit keeps the voltage drop across the regulator to a safe low level.

Although not shown, the unregulated source of the 2500 volts plays a part in the protective process. It should be current limited somewhere beyond the 300-mA full load current, and the rated 3.5 amperes of the LT317A. However, this provision usually will be present whether or not by intent. Because of economics, it is likely that a transformer will saturate or capacitors will run out of energy storage before currents greatly in excess of 300 mA can be supplied to a short-circuited load. Note that the filament transformer must be rated to withstand high voltage.

A practical 100-watt, off-line power supply

The 100-watt, off-line supply shown in Fig. 4-4 represents a practical blend of techniques and devices evolved following the heyday of the 20-kHz switcher. This easy-to-build regulator embodies the following features:

- Off-line operation with electrically isolated DC output
- Five-pin current-mode controller provides fast line regulation, as well as tight load regulation
- Optoisolator in voltage feedback loop preserves isolation established by output transformer
- 40-kHz switching frequency reduces size of magnetics and filter capacitor relative to 20-kHz designs, yet, this rate is low enough so that little trouble should be encountered with parasitic reactances in components and wiring
- Parts count is relatively low and tolerances are not crucial
- The flyback circuit, in conjunction with the cascoded power MOSFET switch, and the Schottky output diode, lead to efficient operation over a wide load range; at full load (20 amps), the efficiency approaches 80%, it is 75% or greater for loads between 4 and 20 amperes
- A wide AC line-voltage range is accommodated easily—rated performance prevails throughout the 90 V to 140 V range of the 60-Hz line. (Circuits of this type are often specified to operate well from from a DC line, or from higher than 60 Hz lines. Operation at 50 Hz or 25 Hz would probably call for a larger filter capacitor following the rectifier bridge.)

The admonition to be cautious of lethal line voltage is not peculiar to this power supply—it relates similarly to any off-line power supply that does not use a 60-Hz isolation transformer at the AC input. As part of this precaution, do not use grounded test instruments to perform measurements within the shaded area. During breadboarding, or during subsequent measurements, it is best to insert a temporary isolating transformer between the line and the supply. Those with any experience in servicing TV sets, or modern radios, will be well aware of this precaution. Once the supply is operating in its intended environment, there is no danger of ground conflicts or inadvertent shorts to the AC line. The electrical isolation between the AC and DC output sections of the supply provides this safety feature.

The power MOSFET switch (MTH7N50) is cascode connected to the bipolar switch within the LT1071 controller. This is an inversion of the usual BIMOS arrangement where a common-base bipolar transistor is driven by a MOSFET. In a way, however, a similar overall objective is achieved—the switching of higher voltage is possible than could be handled by the internal switching transistor alone. The internal switching transistor incorporates an antisaturation circuit; this enables it to switch 40 kHz very efficiently. The cascode arrangement also shifts the heat-removal burden from the controller to the power MOSFET.

Although the LT1071 has its own internal error amplifier, it is not used to sense the sampled DC output voltage in this application. Rather, the output from the external error-voltage processing system (the LT1006 op amp and the 4N28 optoisolator) is applied directly to the comparator at pin V_C. This is done because more stable operation with respect to temperature obtains if the optoisolator follows, rather than precedes error-signal amplification.

From the statement given regarding the internal error amplifier of the LT1071, it would appear that the FB pin need not have any circuit connection. Yet, it is seen that the FB pin connects to a resistance-divider network. The purpose of this is to prevent erratic operation at very low line voltage; under such conditions, the internal error amplifier is able to inhibit operation altogether as consequence of this connection. In normal operation over the 90- to 140-volt range of AC line voltage, the internal error amplifier does not participate in power supply operation.

On the other hand, Q6 and its associated components provide current limiting when the output of the supply is shorted. Although internal current limiting occurs within the LT1071, the drastic abuse imposed by high line voltage and a short-circuited load justify this additional protection. DC output current is limited to about 2 amperes if a short circuit should occur at the output.

At this time, it is necessary to point out the source of positive potential for not only the collector of Q6, but the collector of Q4, and the collector of the npn light-actuated transistor within the optoisolator. This source is the V_c pin of the LT1071 controller, which bears a small positive voltage, in addition to being an input to the comparator.

An interesting feature of this power supply is the transformer secondary winding and the MUR120 half-wave rectifier for powering the LT1071 during operation. What happens is that the LT1071 initially receives its DC operating voltage via the 47 k dropping resistor from the main DC supply. Following a slow start period, normal operation commences, whereupon the major portion of DC operating current needed by the LT1071 derives from the rectified 40 kHz voltage. This technique greatly reduces the power dissipation that would otherwise continue in the 47 k dropping resistance.

Q4 plays a salient role in soft starting the power supply. When the supply is first turned on, Q4 is in its conducting state, holding the V_c pin in a biased-off

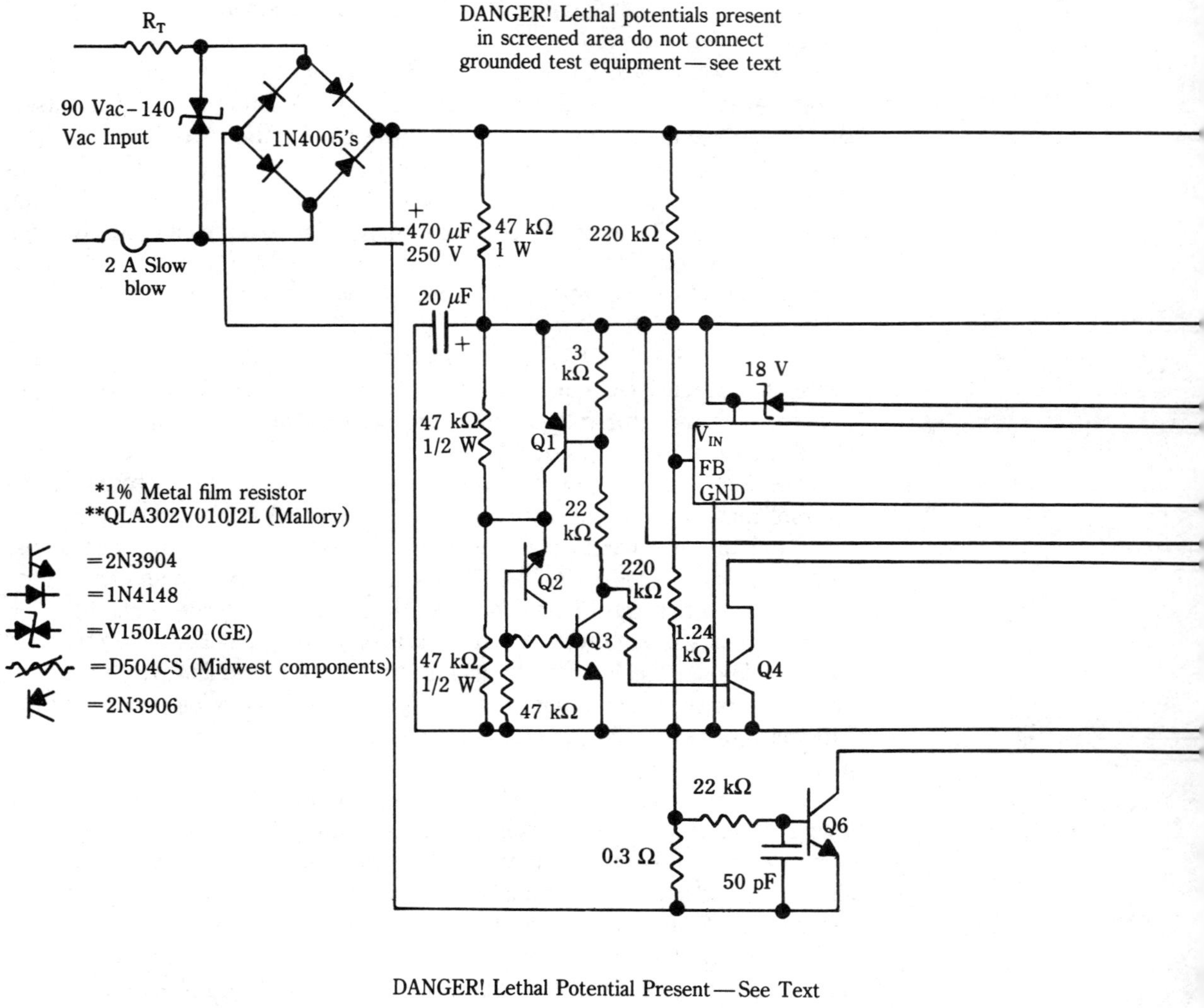

4-4 Easy to build, high performance, 100-watt switching regulator.

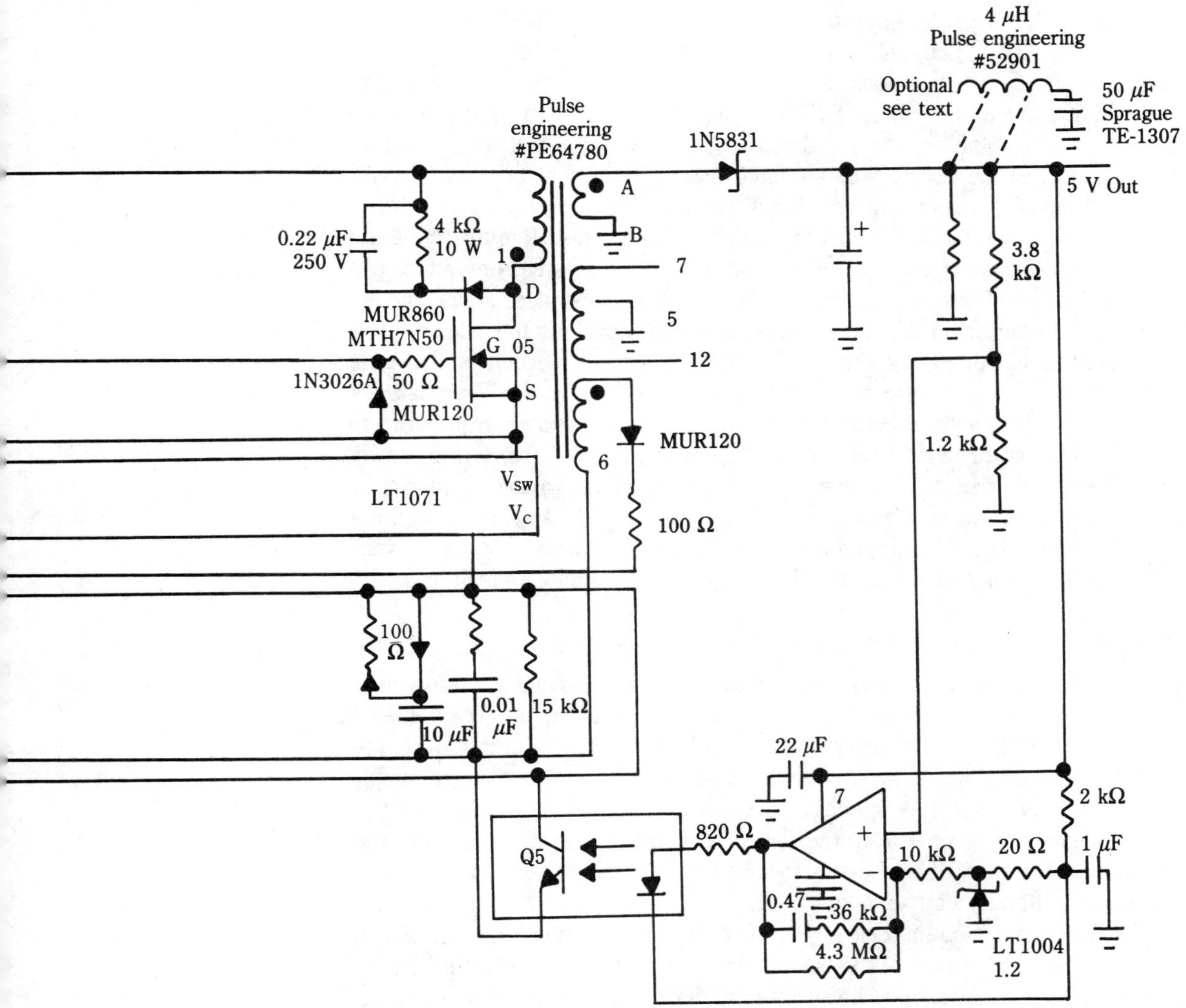
4 μH
Pulse engineering
#52901
Optional
see text
50 μF
Sprague
TE-1307
Pulse
engineering
#PE64780
1N5831
A
B
5 V Out
0.22 μF
250 V
4 kΩ
10 W
1
D
7
5
12
3.8
kΩ
MUR860
MTH7N50
G 05
1N3026A
50 Ω
S
MUR120
MUR120
6
1.2 kΩ
LT1071
V_{SW}
V_C
100 Ω
100
Ω
0.01
μF
15 kΩ
10 μF
22 μF
7
2 kΩ
Q5
820 Ω
10 kΩ
20 Ω
1 μF
0.47
36 kΩ
4.3 MΩ
LT1004
1.2

condition so that the LT1071 produces no switching action. Note that Q4 is controlled by the Q1, Q2, and Q3 combination. This circuitry behaves in a manner similar to an SCR—it is initially in a nonconductive state, but once caused to conduct, it remains on. Q1 and Q3 are the simulated active elements of an equivalent SCR; Q2 functions as a Zener diode to impart hysteresis to the transitional states of the circuit. Keep in mind that the overall function of Q1, Q2, and Q3 is simply to control the conductive state of Q4.

Now, see what happens when the supply is initially placed in operation. As was previously pointed out, the LT1071 at first receives its DC operating voltage via the 47 k dropping resistance. However, this resistance, in conjunction with the 20 microfarad capacitor imposes a time constant that slows down attainment of full operating voltage. During the greater portion of this time delay, Q1 and Q3 are nonconductive, whereas Q4 is conductive. With passage of time, the rising voltage across the Q1, Q3 pair causes these transistors to turn on, thereby turning Q4 off. The 10-microfarad diode network then slowly allows pin V_C to rise to its nominal operating potential. In this sequence of events, two things have been accomplished—the LT1071 has been spared the burden of being placed in operation prior to attainment of full DC operating voltage; also, transients have been avoided by gradual application of load power.

Alternate circuits for resonant-mode supplies

In chapter 1, the resonant-mode power supply was discussed and a practical example was described. It turns out that there are other circuit arrangements for shock exciting a tank circuit and extracting power from it. In Fig. 4-5a, the simplified circuit depicts the technique used in the resonant-mode supply described in chapter 2. It is a popular design approach and its low-loss operation at high frequencies stems from the fact that switching occurs at zero-current intervals. Accordingly, the tendency toward simultaneous current and voltage in the power switch is neatly circumvented.

The resonant-mode circuit of Fig. 4-5b is the half-bridge version of Fig. 4-5a. The two power switches in conjunction with the two circuit-return diodes enables push-pull operation to be achieved without the inconvenience of center-tapping the primary winding of the output transformer. The two capacitors at the input portion of the circuit are the filter capacitors associated with an input bridge rectifier/voltage doubler commonly used with off-line power supplies. In this circuit, the two power switches operate in opposite cadence—one is on while the other is off. The primary of the output transformer is thereby operated in a symmetrical bipolar mode. The nice thing about the half-bridge circuit is that each power switch is impressed with the DC operating voltage rather than twice this value as is true in ordinary push-pull circuits.

A requirement common to both the single-ended switching circuit and the half-bridge circuit is that the primary inductance of the output transformer must be sufficiently high so that its presence does not appreciably alter the LC series-resonant frequency. Note also that the diodes of Fig. 4-5b might or might not represent the internal body diodes of power MOSFETs. When power

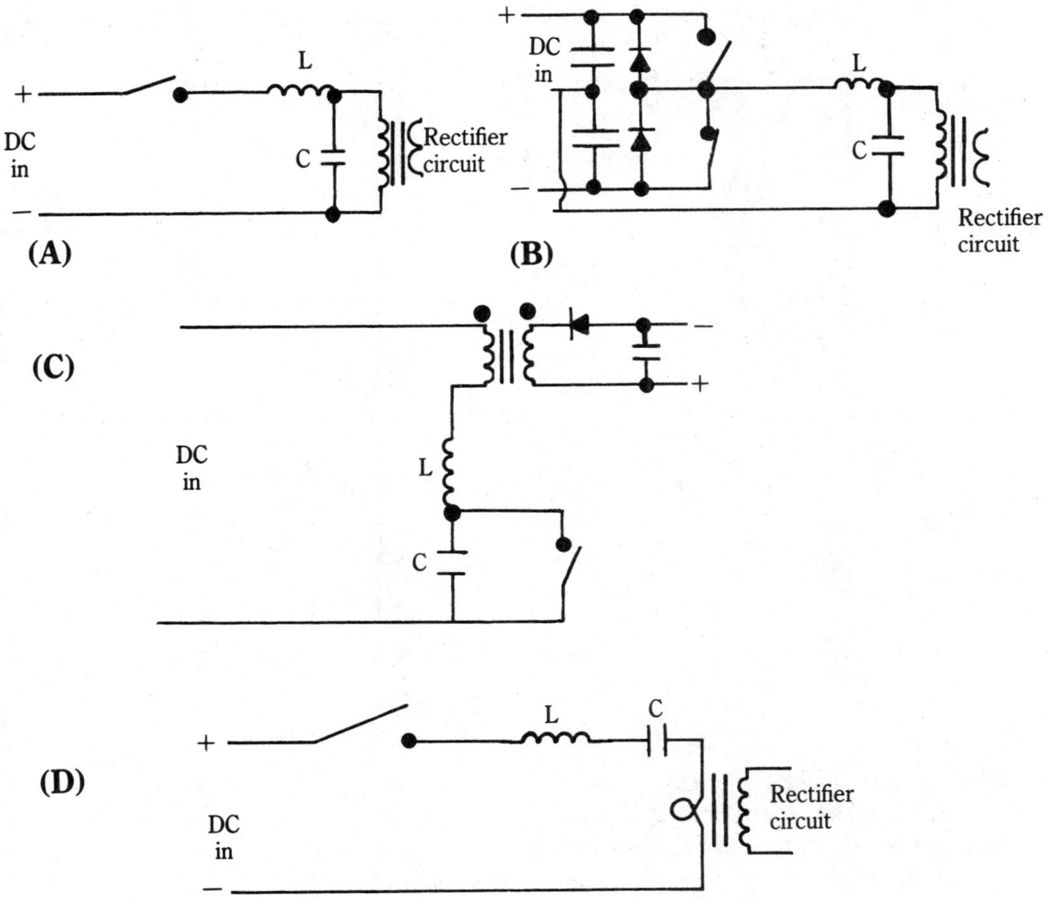

4-5 Variations in resonant-mode topographies.

MOSFETs are employed at higher frequencies, the scheme of Fig. 4-6 is desirable to inactivate the body diodes during portions of the cycle. This is done because the body diodes have relatively slow reverse-recovery times.

The resonant-mode switching arrangement of Fig. 4-5c is commonly used with flyback circuits. In this application, switching transitions occur at times of zero voltage across the power switch. This contrasts with the situations in circuits 4-5a, b, and d in which switching transitions occur at times of zero current through the switch. The net result in both cases is, however, no simultaneous voltage and current in the switch, and therefore, very efficient operation at high switching rates.

The switching circuit of Fig. 4-5d is, in a manner of speaking, the dual of circuit a; in circuit d the primary inductance of the output transformer should be as low as possible—it should be much lower than resonating inductance L. This is so the transformer primary will not appreciably affect the resonance established by L and C. Nonetheless, a variant of this circuit dispenses with inductor

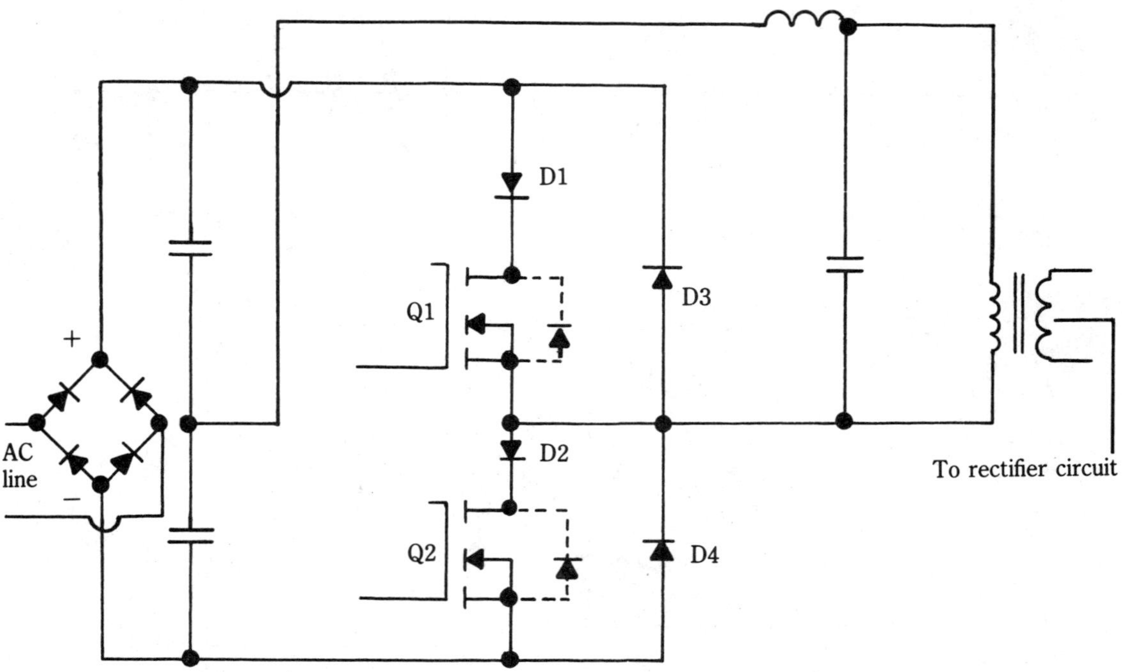

4-6 Resonant-mode, half-bridge output stage for high-frequency operation.

L and uses C to resonate the leakage inductance at the primary of the transformer. (It should be realized that resonating can be made to take place in the secondary side of the transformer, instead of the primary. These circuit variations often form the basis of manufacturer's claims to proprietary designs with superior performance. Most likely, they are the result of inevitable leakage inductance in practical transformers; it is, however, to the maker's credit that "lemonade was made from lemons.")

Circuits c and d of Fig. 4-5 act as constant-current sources, whereas circuits a and b act as constant-voltage sources. This is worth keeping in mind for special applications. The flyback switcher is one application where a current source is appropriate; others are capacitive loads, battery chargers, electroplating, and fluorescent lamps. In most electronic applications, the use of voltage source switching circuits works best.

The circuit shown in Fig. 4-6 sometimes causes confusion; its operation is essentially similar to that of the arrangement of Fig. 4-5b. It happens that the reverse recovery time of the internal body diode of ordinary power MOSFETs is relatively slow. To advantageously exploit the high-frequency capabilities of power MOSFETs and the resonant operational mode, the body diode should not be allowed to function as a freewheeling, or current-return diode. Fortunately, the body diode can be made electrically inactive by blocking its current path with an isolating diode. D1 and D2 are the isolating diodes (often Schottky diodes) in Fig. 4-6. D3 and D4 are external fast-recovery diodes that provide the current-return paths. Because of the circuit arrangement, D4 provides the current-return path when Q1 turns off. D3 provides the current-return path when Q2 turns off.

This scheme is not geared directly to the resonant-mode circuit; it is seen more often with resonant-mode supplies than with those operating on the PWM principle because of the higher switching rates generally encountered. Below some frequency, say 150 kHz, the losses in the isolating diodes might not be sufficiently overcome by the gain in switching efficiency; accordingly this is strictly a high-frequency technique.

Determining tank-circuit values in the resonant-mode power supply

The basic topography of the arrangement shown in Fig. 4-7 is used in many resonant-mode power supplies. A natural question arises regarding the values of the inductor, L, and the capacitor, C. It is obvious that series resonance can obtain from an infinite number of combinations of L and C. It is known that high-Q in series-resonant circuits stems from high inductance and low capacitance for any given resonant frequency. It is also known that the power that can be extracted from the capacitor diminishes with decreasing capacitance. A third

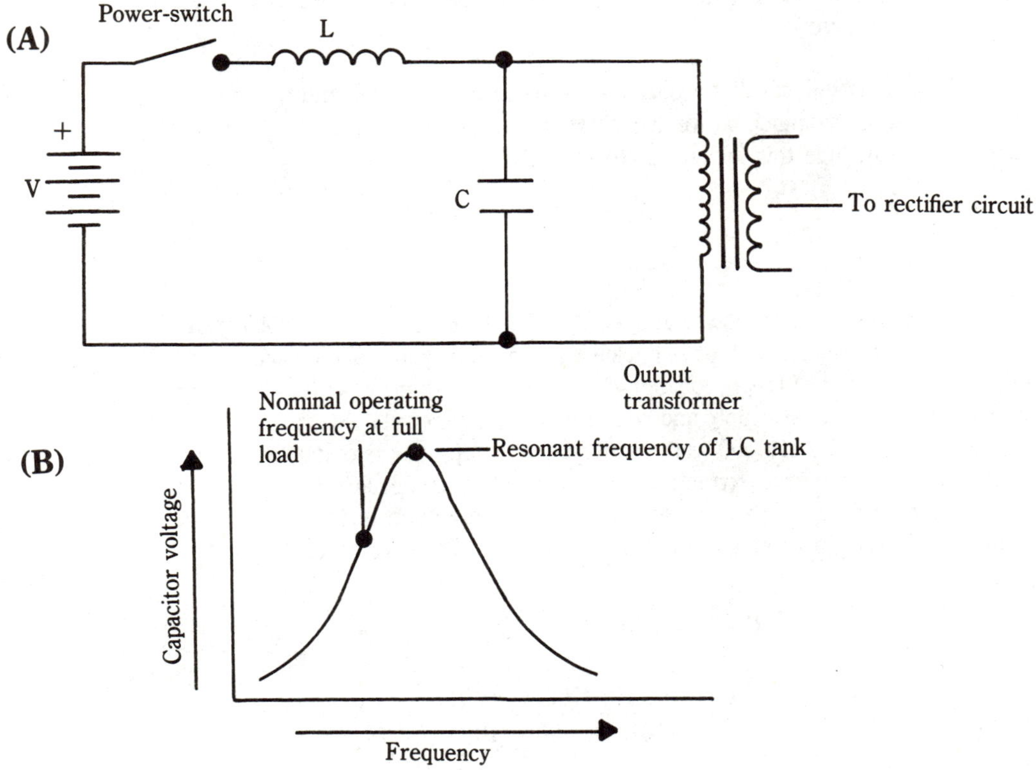

4-7 Basic ideas for determination of LC tank circuit values.

noteworthy fact is that low-Q resonance causes high peak currents in the power switch.

From this information, it is evident that there must be some optimum Q or range of Q's best suited to practical power supplies. Although empirical evaluation is probably necessary for selection of the best inductor-capacitor combination, it would be very helpful to be able to calculate a reasonable combination from which to commence experimentation. This could save not only time, but semiconductors as well.

It turns out that such a calculation can best be found from the nature of energy storage in capacitors, rather from consideration of the somewhat elusive operating-Q of the resonant tank. Of course, once the calculation is made, you can deduce relevant statements about Q. Reference to Fig. 4-7 will prove helpful.

From the physics of energy storage in capacitors, you have

$$P = \frac{CV^2}{2} \times f$$

where: P is the power in watts available from the capacitor
C is the value of the capacitor in farads
V is the peak voltage developed across the capacitor
f is the rate at which the capacitor is replenished with charge, i.e., frequency.

This formula tells us that available power increases with C and f, and even more so with V. Inasmuch as we are interested in the value of C, it serves our purpose to rearrange this formula as follows:

$$C = \frac{2P}{V^2 \times f}$$

In the resonant-mode power supply, f can be chosen to be about 80% of the actual resonant frequency. This is because normal operation takes place on the low-frequency skirt of the resonance curve. Assume some easy to work with numbers for a practical supply and see how things work out:

Let V = 50 volts in a 100 watt resonant-mode supply employing series resonance, in which power is extracted from the resonating capacitor by means of a step-down transformer. From a number of considerations, it is decided that the nominal operating frequency at full load is 500 kHz. Then we have:

$$C = \frac{2P}{V^2 \times f}$$

$$= \frac{(2)\,(100)}{(50)^2 \times 500 \times 10^3}$$

$$= \frac{200}{2500 \times 500 \times 10^3}$$

C = 0.16 microfarad

Next, from the classicial formula for series resonance,

$$f = \frac{1}{2\pi\sqrt{LC}}$$

the value of inductor, L, can be expressed as:

$$L = \frac{1}{C \times 4 \times \pi^2 \times f^2}$$

However, before using this formula, recall that the operating frequency at full load and at the specified 50-volts input will be about 80% of the LC resonant frequency. Therefore, the value of f now used will be

$$\frac{500 \text{ kHz}}{0.8} = 625 \text{ kHz}$$

Solving for L:

$$L = \frac{1}{0.16 \times 10^{-6} \times 4 \times (3.14)^2 \times (625{,}000)^2}$$

$$= \frac{1}{0.16 \times 4 \times 9.86 \times 391 \times 10^3}$$

$$= 0.41 \text{ microhenry}$$

Note that these calculations have circumvented dealing with Q directly. However, we can nonetheless be guided by the fact that a higher Q resonant tank (higher L, lower C than calculated) will tend to decrease the amount of power that can be extracted. Conversely, a lower Q resonant tank (lower L, higher C than calculated) will tend to increase the peak current that the power switch will be exposed to.

It is obvious that your chance of ballparking the resonant tank by merely establishing resonance from an arbitrarily selected inductor and capacitor is slim.

The Cuk converter

The basic Ćuk converter shown in Fig. 4-8 has the potential for high efficiency, and, in principle, can operate with very little ripple injected into the line or into the load. A corollary of this is that this circuit should be relatively free of RFI and EMI. It naturally inverts the polarity of the input voltage source, and can be operated to deliver either a higher or a lower load voltage than the input voltage.

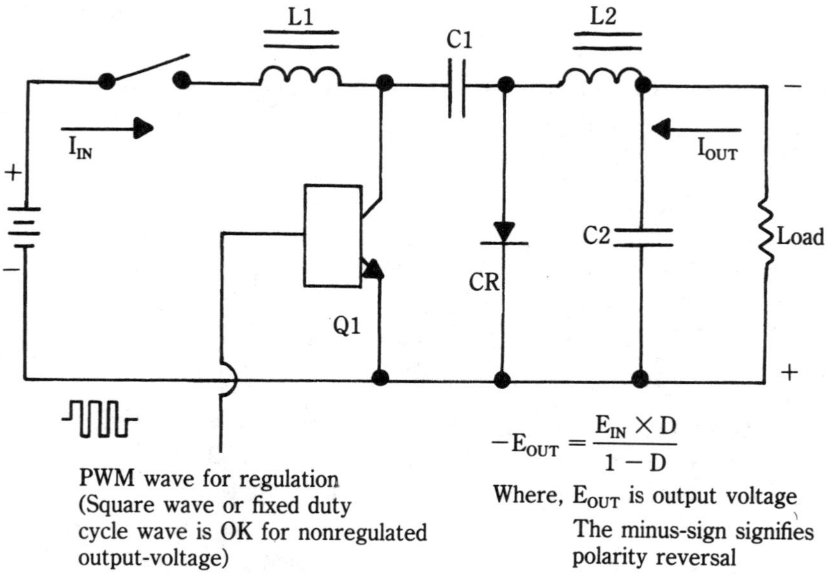

$$-E_{OUT} = \frac{E_{IN} \times D}{1 - D}$$

Where, E_{OUT} is output voltage
The minus-sign signifies polarity reversal
E_{IN} is input voltage
D is the duty cycle

4-8 The basic Ćuk converter.

The popularity of the Ćuk converter has waxed and waned since its development at the California Institute of Technology in 1976. A practical difficulty is that near ideal components are required for reasonably superior performance over other converter topologies. This pertains especially to the capacitor C_1, which must carry a heavy ripple current, and must have a low effective series resistance. However, recent improvements in both active and passive components, together with the trend to higher switching rates brings about some practical compromises that compel renewed consideration of the Ćuk converter. For example, the availability of low R_D power MOSFETs, polypropylene capacitors (nonelectrolytic), Schottky diodes, and low-loss, high-frequency core materials are just what the doctor ordered. Also, certain circuit modifications, such as the mutual coupling scheme depicted in Fig. 4-9, have further refined the performance of practical circuits.

The salient operating features of this circuit is that it tends to consume a steady current from the input source, and deliver a steady current to the load with minimal dependence upon brute-force filtering. Load power derives in alternate fashion from energy stored in inductor L1 and energy stored in capacitor C1. This suggests resonant-mode or quasiresonant operation, but the Ćuk converter is not so classified. The switching transistor apportions energy alternately from magnetic and electric storage; there is no ongoing "natural" LC oscillation and sine waves are not involved.

Suppose that switch SW is closed, connecting the DC source to the converter circuit. Current will flow in the L1, C1, CR branch; in so doing, C1

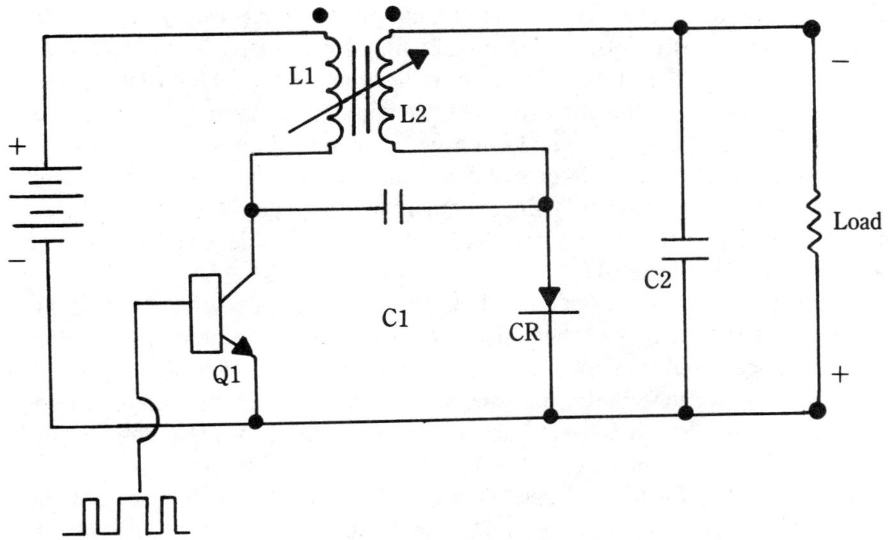

4-9 Modified `Cuk converter.

becomes charged to input voltage, E_{IN}. (During this interval, transistor Q1 was "open.") Next, assume Q1 is keyed into its conductive state; two things happen: C1 discharges, forcing current through L2 and C2 and the load. Also, input current finds a path to flow through the charge up L1. When Q1 again opens, the stored energy in L1 sends current through C1 and CR, thereby recharging C1. CR now also completes the path for the continuing flow of L2 current. With the next turn on of Q2, the cycle repeats. Keep in mind that current directions are as indicated by the arrows, and output-voltage polarity is inverted.

Current-mode regulation controllers

The IC controller for swtich-mode power supplies represented a quantum jump in regulator technology. For a long time, these ICs were tailored for use with pulse-width modulation circuits. More recently, control ICs for dedicated application in resonant-mode supplies have made their debut. The implementation of both modes of control ICs greatly reduced the parts count of switch-mode power supplies, thereby enhancing their reliability, improving overall performance, and reducing cost. Additionally, circuit functions promoting operating convenience and protection to both power supply and load became easy to implement. Thus, a typical control IC might, in addition to basic input and output provisions, have pins to provide such auxiliary functions as soft start, selectable drive option (single ended or push pull), shutdown, oscillator frequency (via external RC network) synchronization, reference-voltage output, dead time, remote sense, under- and overvoltage sense and shutdown, etc.

However, virtually all of these control ICs were voltage-mode types, whether or not this was actually specified. By "voltage-mode" is meant the sampling of the regulator output voltage as means of producing appropriate servoaction to maintain the output voltage at a constant level. In PWM supplies, the servoaction, or error-correction process, is accomplished by varying the duty cycle of the output pulses delivered by the power switch. These pulses are then integrated by the LC filter so that they become nearly constant DC. In resonant-mode supplies, the corrective servoaction derives from a constant-duration pulse that is frequency modulated by the supply's output voltage; these pulses emerging from the power switch are impressed on a resonant tank circuit that then provides an amplitude dependent upon the frequency of the pulses. As with the PWM regulator, an output filter integrates the pulses so that the load sees essentially constant DC.

Although the voltage-mode controller has rendered yeoman service, it has all along been plagued by certain difficulties. Consider, for example, the matter of line-voltage regulation. A change in line voltage will tend to have its effect regulated out because the accompanying tendency for output voltage change will be regulated out. However, such correction cannot take place immediately because of the delay imposed by the LC output filter. Therefore, fast line transients cannot be effectively regulated out. Instead, a substantial portion of line transients can pass through the power supply and appear at the load. This is because of the nonideal characteristics of the output filter—distributed capacitance in the inductor, and effective series resistance and inductance in the capacitor.

Another shortcoming of the traditional voltage-mode regulating system is the matter of stability. This, to be sure, has been no problem in successful designs, but only because of less than desired gain in the error amplifier. The amount of gain allowed depends heavily on good phase/gain compensation in the feedback loop, but is basically limited because of the output filter. The 40 dB per decade roll off imposed by the typical output filter tends to make compensation difficult if the supply is to be used under a wide variety of operating conditions. This situation was accepted mainly because no easy remedy was in sight.

Yet another negative feature of voltage-mode regulation is the matter of protective current limiting. It can be appreciated that, again because of the delay of the output filter, the attenuation of an over-current condition in the power switch cannot instantaneously take effect. Thus, protection is not as effective as would otherwise be the case.

As you might well expect, the solution to these several deficiencies of voltage-mode regulation is the use of current-mode operation. However, it should be realized that the current-mode scheme is usually employed in addition to, rather than simply in place of, voltage-mode operation. Thus, a current-mode control IC generally provides for two feedback paths—one for sensing a portion of the power-switch current, and the other for sensing a sample of the output voltage. The main task of the current-sensing feedback is to instantaneously regulate against power-line transients and to provide cycle-by-cycle current limiting in the power switch. The main task of the voltage-sensing feedback is the traditional one of regulating the output voltage of the supply. These two feedback loops work together in such a way that the output filter no longer imposes severe compensation problems, and much higher gain can be used in the error amplifier than when voltage-mode operation is used alone. This tends to produce closer regulation, a lower output impedance, and less ripple.

The close similarity but significant differences between voltage-mode and current-mode regulators are illustrated by the simplified block diagrams of Fig. 4-10. Note that the PWM comparator in both modes is supplied with a triangular (or sawtooth) wave, but from different sources. Also, in the current-mode system, there is a feedback loop within a feedback loop. In the current-mode arrangement, R_s is a very low resistance so as not to impair efficiency, but to produce a voltage drop representing the current ramp through the power switch. Alternately, a current transformer is often used for the sensing function.

An instructive insight into the nature of the current-mode regulator can be attained by considering the triangular wave at the input of the PWM comparator in the voltage-mode circuit. The amplitude of this wave is kept constant under normal operation. Assume, however, that its amplitude changed during opera-

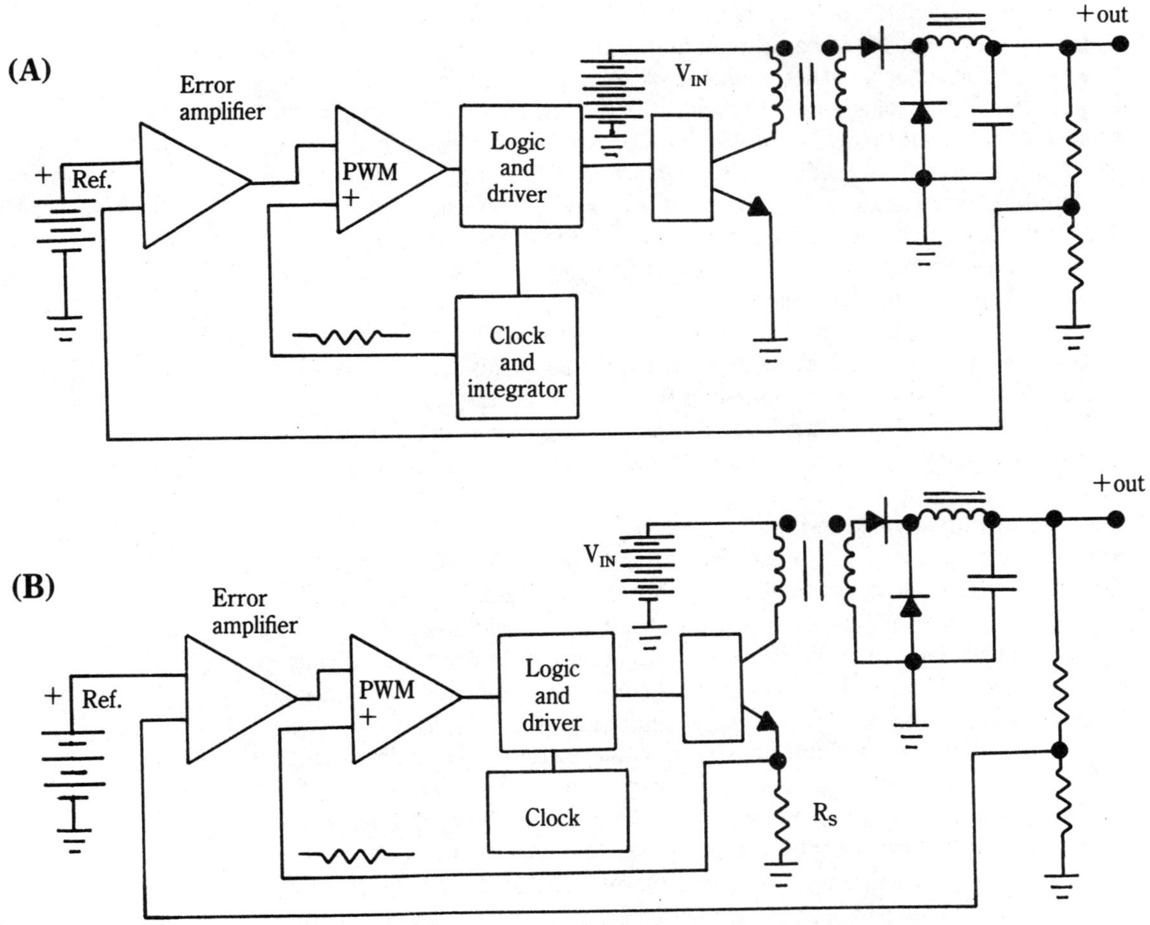

4-10 Voltage mode vs current mode PWM regulators.

tion of the regulator. What would happen to the output voltage? On first thought, it would seem that a change in the triangular-wave amplitude would be accompanied by a resultant change in the output voltage of the regulator. Observe, however, the servoaction of the error amplifier/voltage reference circuit remains intact; it continues to adjust its output so that a null exists at its input. This being the case, the changed triangular wave will not affect the output voltage any more than an approximately similar percentage change of some other parameter, such as V_{in}.

This shouldn't be surprising considering what the feedback loop has been intended to do. Indeed, unless you are splitting hairs, it is essentially true that only a change in the output sampling potentiometer or a change in the reference voltage can materially affect the output voltage. The higher the gain of the error amplifier, the more nearly this statement applies in practice as well as in principle.

Now that it is realized that a constant-amplitude triangular wave is not necessary to maintain constant output voltage, it becomes relevant to consider the situation in the current-mode regulator. Here, the PWM comparator works in a similar manner to the PWM comparator in the voltage-mode system. What happens is that the error amplifier seeks a null from a greater variety of circumstances than in the voltage-mode system; the fact that it seeks the null is the significant factor, for voltage regulation of the output is thereby accomplished just as it is in the voltage-mode regulator. In addition, however, variations of V_{IN} are almost instantly regulated out because the internal (current sensing) feedback loop does not include the delaying action of the output filter. This statement, admittedly, might not be easy to confirm, for an additional subtlety in circuit performance must be taken into account.

When V_{IN} changes, say because of a line-voltage transient, the current-ramp sensed at R_S changes its slope. It is this slope change that changes the duty cycle in such a way as to counteract the tendency of the transient to alter the output voltage. Bear in mind that this slope-correction phenomenon takes place quickly—long before the output voltage-sensing servoaction can get into the act.

In addition to crisping up the PWM wavetrain, the logic and driver function in the block diagrams of Fig. 4-10 provide such "bells and whistles" as dead time, duty cycle limiting, double-pulse suppression, push-pull drive, controlled inhibit, automatic shutdown, undervoltage lockout, and synchronization.

Examples of current-mode ICs are shown in Figs. 4-11 and 4-12. Although these ICs contain numerous protective and convenience functions, they are in essence similar to the simplified arrangement of Fig. 4-10b. In all cases, it will be seen that, despite the "current mode" nomenclature, two feedback loops are present, one for sensing the current ramp through the power switch, the other for sensing the sampled output voltage. In all cases, the sawtooth generator (oscillator plus integrator) of ordinary "voltage-mode" IC controllers is replaced by the sampled ramp through the power switch. The net result is that these ICs simultaneously regulate against changes in the AC power line, and against load changes.

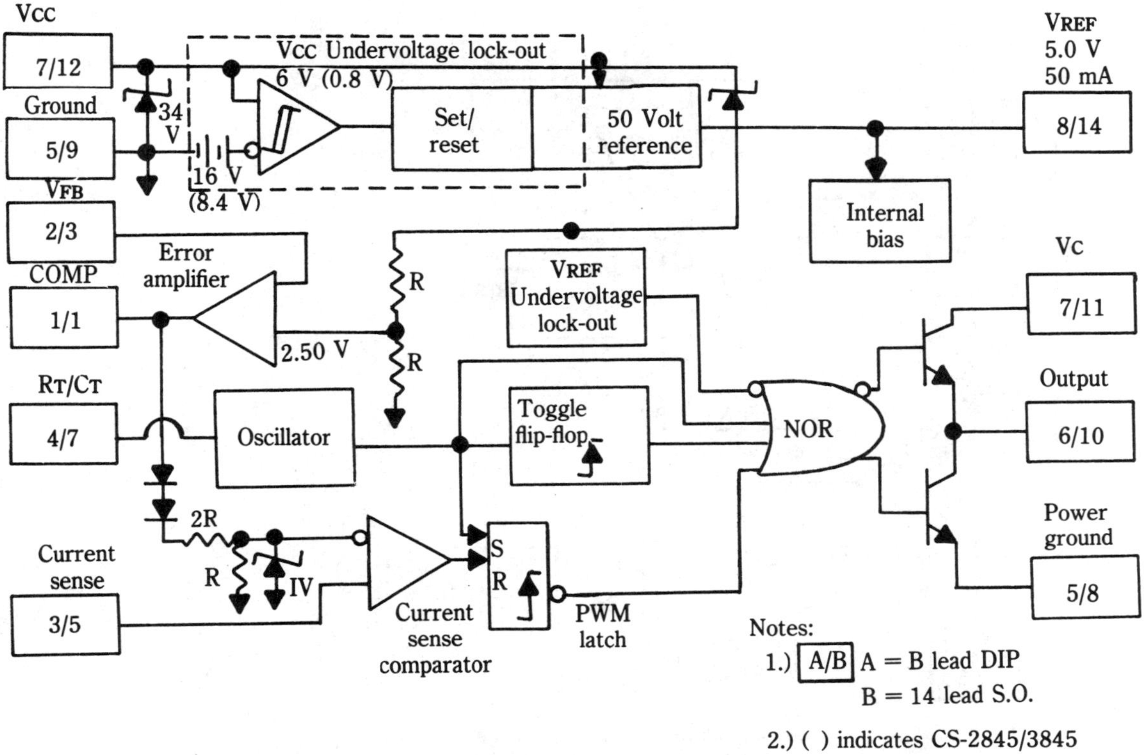

4-11 Block diagram of current mode PWM controller IC.

In summary, the features of the current-mode control technique are as follows:

- Virtually instantaneous correction of input voltage changes—transients on the AC power line are much less likely to reach the load than is the case with controllers using voltage-mode sensing only
- Cycle-by-cycle limiting of current confers greater protection to the power switch than can be achieved readily in voltage-mode only control
- The effects of phase shift and roll off in the output filter are greatly reduced. This relaxes frequency-compensation problems in the error amplifier, permitting higher loop gain, better stability, and tighter regulation; often, this also leads to reduced output ripple and noise.
- Current-mode control appears to work especially well for two important circuitries, resonant-mode switchers, and active harmonic filters. The latter circuit appears destined for increased development and application inasmuch as it enables near-unity power factor to be realized from supplies that would ordinarily consume heavy AC line current because of low power factor

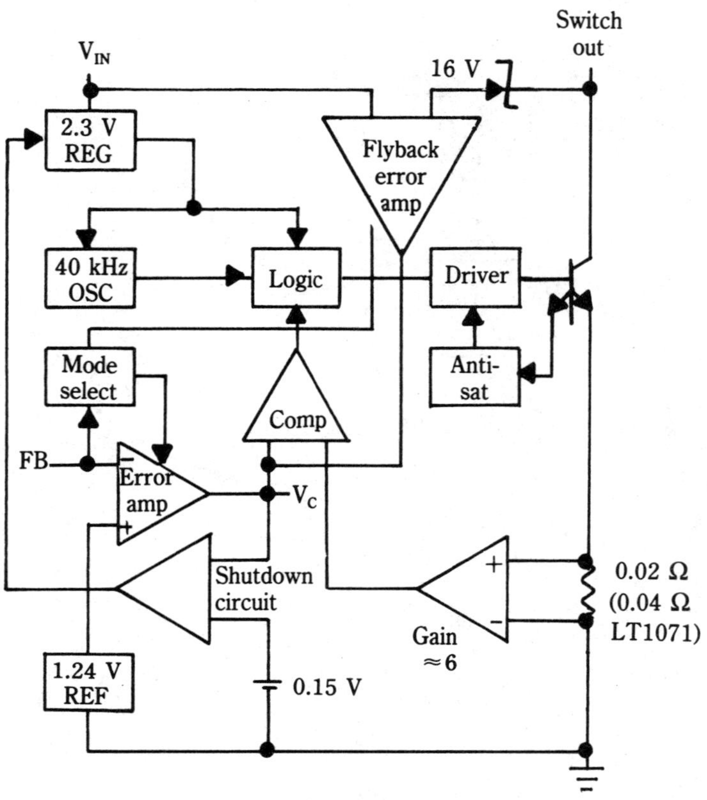

4-12 Block diagram of a complete current mode PWM power supply.

Active harmonic filter for correction of power factor

Correction of the power factor of power supplies is an important function because the AC current consumption of a supply tends to be considerably higher than the value calculated from a budget of its overall dissipative losses. Power factor means the ratio of true power to apparent power, where true power is the input power that (except for dissipative losses) is delivered to the resistive component of the load; apparent power is AC input current times AC input voltage without regard to their phase relationship. True power is measured by a wattmeter; apparent power is calculated from ammeter and voltmeter readings. Power factor is true power/apparent power.

Generally, the AC line voltage and current are not in phase inasmuch as the power supply exhibits reactance. This is not surprising because the typical input circuit comprises a rectifier bridge operating into a filter capacitor. Because of the cyclic charging current of the capacitor, the line current assumes a peaked shape. See Fig. 4-13. It has been overlooked often that a sinusoidal AC voltage with a nonsinusoidal current behaves as if reactance is present with its attendant phase displacement between impressed voltage and consumed current. Typical regulated power supplies do, indeed, contain physical reactances, predominantly the filter capacitor. However, it is important to grasp that the phase displacement prevails whatever the cause of nonsinusoidal line current. A case in point is the nonsinusoidal line currents produced by SCR and triac circuits where no filter capacitor is present.

Phase displacement between applied AC voltage and consumed AC current is tantamount to a less than unity power factor. This, in turn, makes apparent power greater than true power, and is accompanied by needlessly high AC line current. This gives rise to excessive heating and voltage drop in AC wiring, trips circuitbreakers, blows fuses, and overtaxes other line facilities such as transformers. It forces the utility company to generate more power and higher current than would be needed to supply resistive loads. Here, then, is the solution: The power supply must be made to "look" resistive. Traditional techniques of reactance cancelling (resonance) are not easily applicable because of the high harmonic content of the distorted AC current wave. What is needed is a harmonic filter that is also capable of locking AC voltage and AC current in step. Such a situation then will be interpreted by the AC line as a nonreactive, or resistive load, and the operating power will be ideal unity.

Another important consequence of such a harmonic filter is that it also will eliminate a source of interference to other electrical and electronic equipment. It is not uncommon for communications facilities to be plagued by harmonic energy emanating from distorted 60-Hz currents. Fortunately, an effective harmonic filter can be implemented by a modification of the current mode PWM controller.

The simplified block diagram of Fig. 4-14 shows the basic technique where the AC line current can be made to duplicate the sinusoidal shape of the line voltage, and to be in phase with it. The arrangement is essentially a modified current-mode control system, such as is depicted in the bottom drawing of Fig. 4-10. Note the imposition of the functional block labeled "combiner." It allows

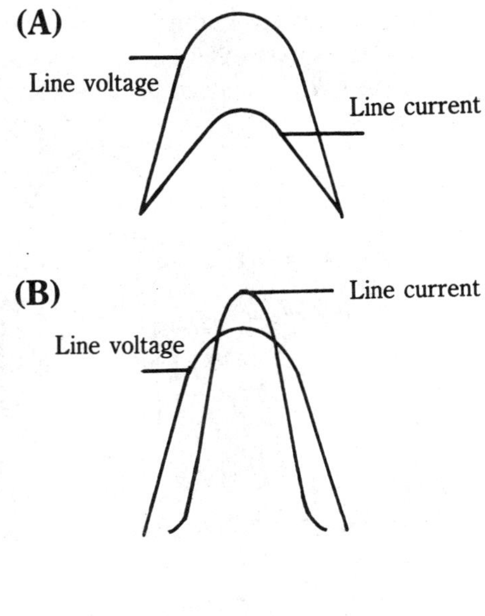

4-13 Half-cycle views of AC line current and voltage relationships.

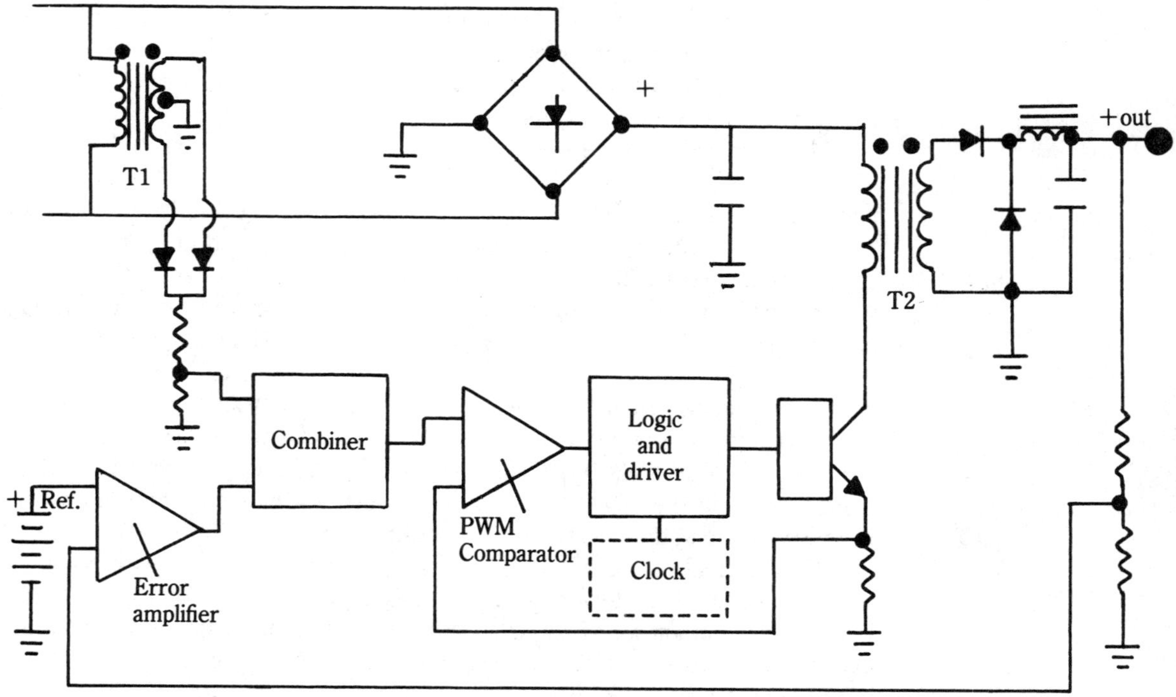

4-14 Basic scheme for correcting power factor.

modulation of the error voltage at the output of the error amplifier by sampled half-sinusoidal voltages derived from the AC line. The output of the PWM comparator is thereby modified so as to cause the smoothed AC line current to follow the line voltage; this action overrides the tendency for the filter capacitor to produce a peaked line current. The restoration of unity power factor operation is illustrated in Fig. 4-15. The AC line now "sees" a resistive load.

Simultaneously with the waveshape restoration of the AC line current, conventional DC output voltage regulation takes place via the voltage-mode feedback to the PWM comparator. This occurs at the operation frequency of the

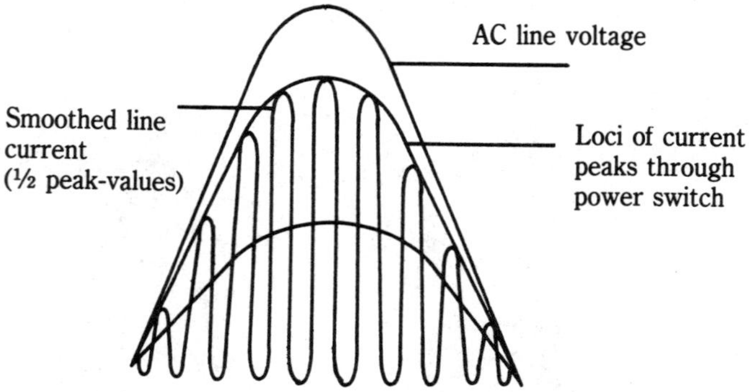

4-15 The restoration of unity power factor operation.

system, say 100 kHz, whereas the power factor correction is a 60 Hz phenomenon; there is no conflict between the two functions.

Certain practical aspects of this scheme merit mention. The small isolation transformer, T1, probably would not be used in a commercial implementation of the harmonic filter. It is depicted as a means of facilitating insight, and also as a desirable component in breadboard experimentation to avoid ground conflicts. The essential idea, in any case, is to sample the AC line voltage (if the transformer is used, heed should be paid to its phasing).

Also, the clock probably would not be necessary; the circuit performs better if it is operated, via an AC feedback loop, as a self-excited system. This is because the triangular waves will then be continuously adjacent to one another—gaps interfere with proper operation. The switching logic is also involved in maintaining this operating condition.

Finally, such a harmonic filter is usually used as a line conditioner ahead of another power supply or inverter, rather than by itself. At first thought it might seem that the overall dissipative losses would lower operating efficiency and defeat the intent of adding the harmonic filter. In actuality, the improvement in power factor, typically from about 0.6 to 0.95, overcomes the degradation of the additional losses, and the effective efficiency of the system is considerably improved.

Switched capacitor power supply

A common denominator of most switch-mode power supplies is the inductor or choke in which energy is stored magnetically and released electrically. Such a technique might seem the most natural approach for the purpose, but it is not without its problems and difficulties. You have to contend with undesirable, and sometimes destructive side effects of the sought energy transformation. Included in the list of shortcomings of inductors are such ills as saturation, ringing, and phase-shift effects in the feedback loop. Additionally, cost, bulk, and leakage fields often loom up as negative factors. It does not matter much whether the source of electromagnetic storage is a separate inductor or the primary of a transformer. Certainly, many have fantasized how much better things would be if the magnetic-core storage device could be dispensed with.

Such wishful thinking has, indeed, materialized in the form of switched-capacitor voltage converters. As might be expected, such circuitries use capacitors, rather than inductors as the energy-storage devices. In a very real sense, such an approach is more straightforward than electromagnetic storage. Capacitors work with charges—already an electrical quantity. Such circuitries are made in the format of dedicated ICs. These can be used for many of the functions provided by the familiar switched-inductor circuits. Thus, you can accomplish voltage transformation, regulation, polarity inversion, synchronization, and paralleling. For practical reasons, current capability tends to be limited to the 10 to 100 milliampere range in the ICs. However, outboard boosters can readily extend current ratings into the ampere region. Similarly, external voltage boosters can provide higher voltage capabilities.

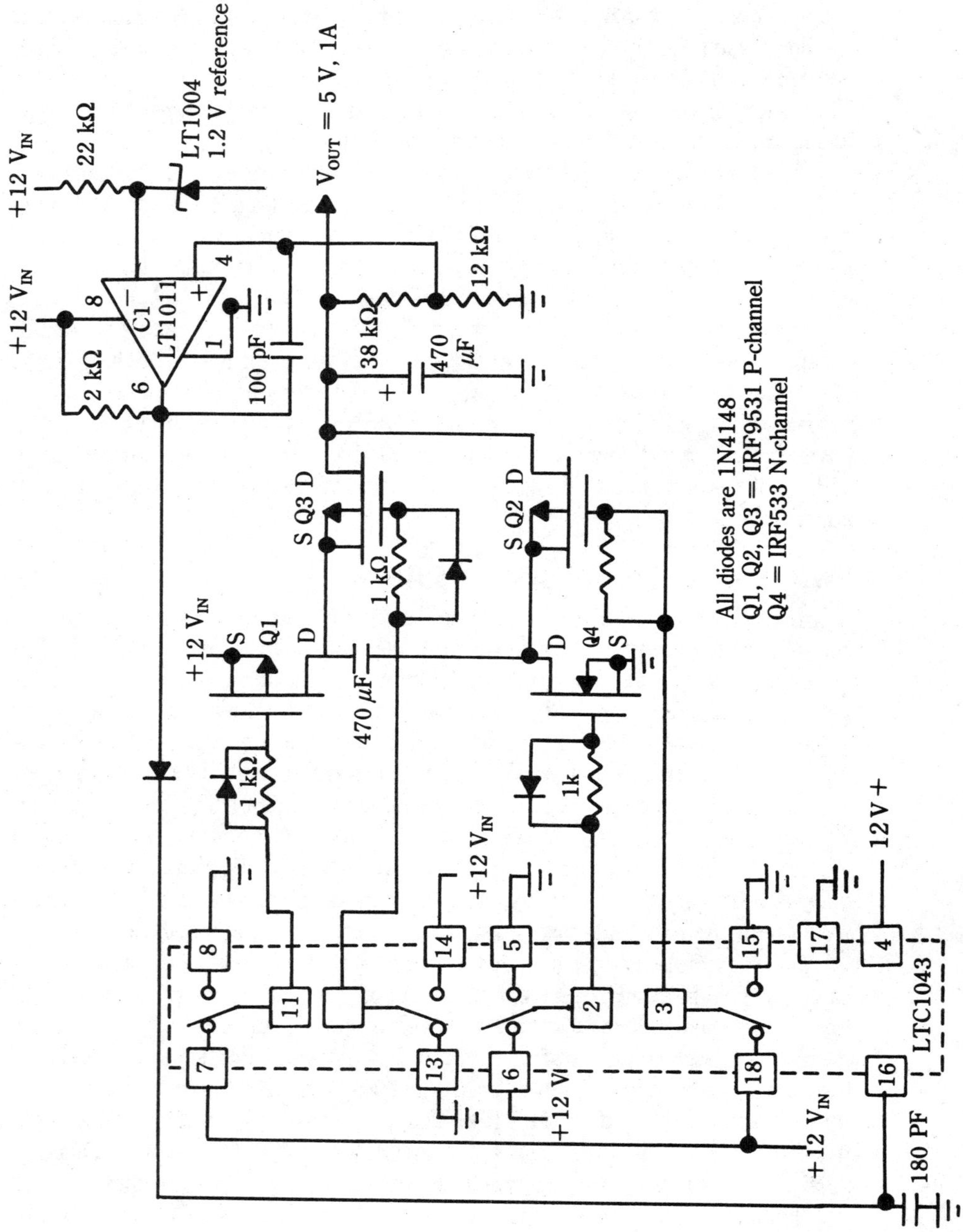

4-16 Five-volt, one-amp switched-capacitor power supply.

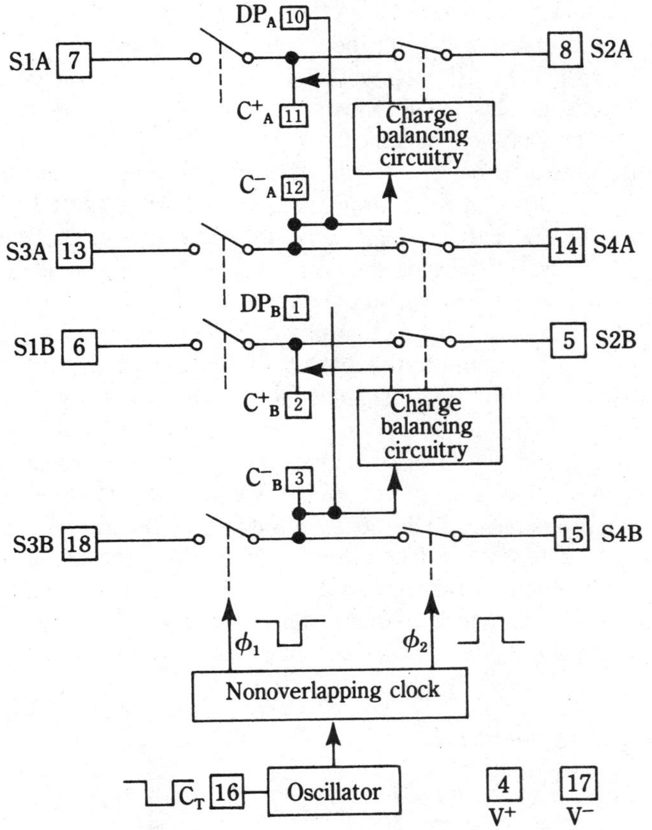

4-17 Architecture of the LTC1043 switching IC.

Figure 4-16 shows an example of such an inductorless switched capacitor power supply. The heart of the circuit is the LTC1043 IC, which is essentially a CMOS switching array. The switching elements are open and closed via a non-overlapping clock, which, in turn, is driven by an RC oscillator. The basic architecture of the LT1043 is shown in Fig. 4-17. The freerunning frequency of the oscillator is approximately 200 kHz, but the frequency can be lowered by connecting capacitance from pin 16 to ground. Moreover, applying a positive voltage to pin 16 will change the duty cycle of the oscillation. Internal charge balancing frees the designer from the effects of stray capacitance. Note that there are two DPDT switching sections in the IC.

This power supply operates by alternately switching the 470 microfarad capacitors from a series to a parallel connection. While in series, the capacitors continue to receive charge while supplying load power; when in parallel, the capacitors undergo partial discharge by the load, but are not replenished from the 12-volt source. The series connection is established when switch elements Q1 and Q2 are turned on; parallel connection of the capacitors occur when Q3 and Q4 are turned on. Keep in mind that Q3 and Q4 are off when Q1 and Q2 are on. By the same token, Q1 and Q2 are off when Q3 and Q4 are on. In other words, there is no overlap, or conflict, between the two circuit states.

Now see how the feedback loop works. Its function is to produce a stabilized 5 volts at the load. The manner in which this is achieved is somewhat different than is commonly encountered, even though the feedback circuit resembles those ordinarily used in regulated power supplies. Insight into the principle employed can be gleaned from the simplified circuits of Fig. 4-18. These circuits are not, strictly speaking, accurate equivalents to the actual circuitry of the supply, itself. However, they are particularly useful in demonstrating the conversion principle involved in the supply.

In Fig. 4-18a, the series-connection phase is depicted. (Assume the RC time constant is large compared to the duration of interest.) Note that the load voltage approaches 6 volts after closure of the switch at time zero. In other words, the large capacitors act as a potential divider.

Next, you see in Fig. 4-18b the situation when the charged capacitors are connected in parallel and separated from the 12-volt source. Now the load simply commences a discharge cycle of these capacitors and the load voltage falls. These alternate operating states take place in the actual power supply, but the feedback loop keeps the excursions from the desired 5-volt output very small. The basic idea involved is that the output voltage can be regulated by appropriately controlling the duration of the series-connected phase of the capacitors.

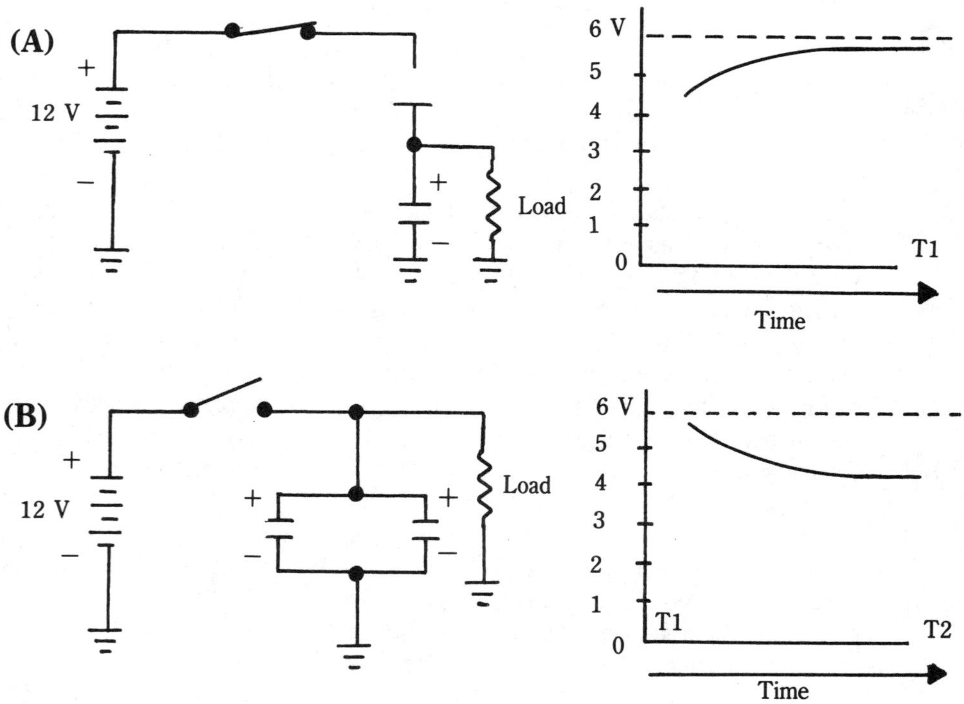

4-18 Simplified circuits showing the alternate load delivery schemes.

Let's divert a moment from our analysis of this particular supply to emphasize a feature inherent in *all* switched-capacitor supplies. Assume you have a 5 microfarad capacitor charged to 20 volts. If this capacitor is placed in parallel with an uncharged 5 microfarad capacitor, it is clear that you then have 10 volts and ten microfarads. The ''knee-jerk'' reaction to this simple situation is that the stored energy has remained constant—that it has only altered its format. A quick look at the arithmetic of the energy interchange brings out the truth of the matter.

The energy stored in a capacitor is

$$\frac{C(V)^2}{2}$$

If the capacitance, C, is in farads, and the potential developed across its terminals is in volts, then the energy content, E, is in joules. Accordingly, let E1 represent the energy stored in the originally charged 5 microfarad capacitor. Let E2 represent the energy stored in the paralleled 5 microfarad capacitors. This is the energy corresponding to 10 volts and 10 microfarads.

$$\text{Then: } E1 = \frac{5 \times 10^{-6}(20)^2}{2}$$

$$= 5 \text{ X } 10^{-6} \times 200$$

$$= 1000 \text{ microjoules or } 1 \text{ millijoule}$$

and:

$$E2 = \frac{10 \times 10^{-6}(10)^2}{2}$$

$$= 10 \times 10^{-6} \times 50$$

$$= 500 \text{ microjoules or } 0.5 \text{ millijoule}$$

What has happened to half of the originally stored energy? This question poses some thought, for no resistance or other form of dissipation was postulated. Interestingly, if a large amount of series resistance had been involved between the two capacitors, and then shorted out when the capacitors attained equal voltage, the available energy would still be substantially 0.5 millijoule. The only effect would be that you would have to wait a bit longer to access energy level E2.

As good an answer as any is that nature has anticipated inevitable losses in the charge redistribution process. The practical ramification for switched-capacitor power supplies is that energy is lost when charges are switched into or out of capacitors. This is true for the best ''low-loss'' dielectrics.

The timing diagrams of Fig. 4-19 illustrate how the feedback loop controls the duty cycle of the alternate series-parallel phases. Keep in mind that the feedback loop performs this duty cycle modulation by influencing the series

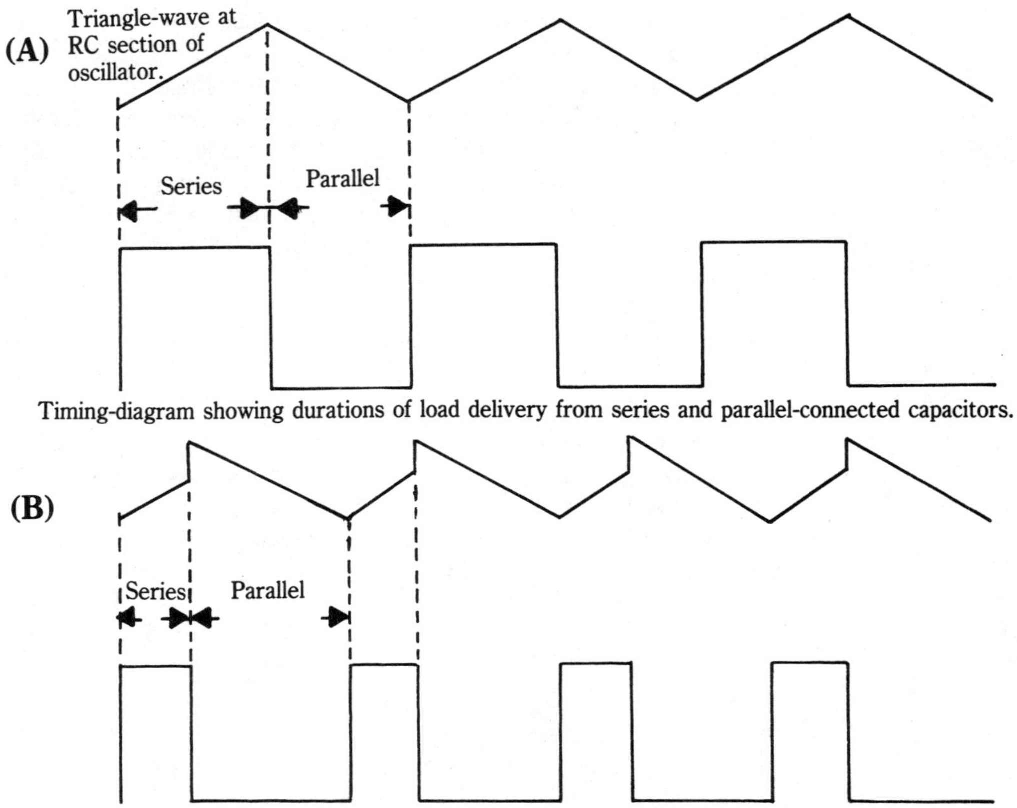

4-19 Timing diagrams showing effect of feedback on load delivery.

phase portion of the cycle alone. In Fig. 4-19a, the triangular wave of the RC oscillator produces a 50% duty cycle in the driven clock. Assume that the upward slopes of the triangular wave correspond to the operational phase in which the load-delivery capacitors are connected in series. This 50% duty cycle occurs as the "natural" operating condition of the oscillator—when the feedback loop is not in place.

In Fig. 4-19b, the output voltage of the supply has attempted to rise above 5 volts and the resultant error signal, a positive voltage step, is applied to pin 16. This interrupts the RC timing of the oscillator; specifically, the series-connection phase is abruptly shortened. This tends to drop the output voltage of the supply. Because of the sensing and servonature of the feedback loop, such interruptions of oscillator RC time constantly occur during the regulation process. Note that if the output voltage attempts to fall below 5 volts, the feedback loop simply allows more time for duration of the series phase of operation. Restoration to the regulated 5-volt output then occurs because of the tendency of the output voltage to "seek" 6 volts in the series phase.

The purpose of the diode ahead of pin 16 is to isolate the error amplifier from the oscillator during the downward slope of the triangular wave. This is in accordance with the objective of controlling only the duration of the upward slope, that is, of the series phase of the operational cycle.

Aside from these unique operational features, it will be seen that the error amplifier, the voltage reference, and the potentiometric-voltage sampler duplicate both form and function of similar feedback components in "conventional" power supplies.

The switching frequency of this supply at full load is about 12 kHz, where it is found that optimum efficiency (about 83%) is attained. At much lower or higher frequencies, switching losses tend to negate other operational advantages in this type of power supply.

Some notes on power darlingtons

The power darlington transistor might have been discussed in chapter 2 along with discrete bipolar transistors; however, certain aspects of the darlington transistor make it a candidate for separate treatment. These devices have been mature for a long time, although asymptotic improvement continues. What is unique about the power darlington is its reputation as a high-beta bipolar transistor—a viewpoint that is at once, both right and misleading. It is not difficult to find schematic diagrams in which the darlington is drawn as a single discrete bipolar transistor. The implication is that all three-terminal devices with base, emitter, and collector electrodes must be essentially the same.

Indeed, for some purposes, such a notion can be considered valid. For example, when a power darlington is used for the series regulator in a linear power supply, practical convenience is well served by treating the device as a discrete bipolar transistor with high current gain. To the purist, however, this analogy might not be acceptable because the driver to a single discrete transistor must overcome a voltage gap of one emitter-base junction, but "sees" two emitter-base junctions at the input of the power darlington. But, this, and a few other nonidentical situations do not seriously detract from likening the darlington to a high-beta discrete transistor when used in series regulators. The four basic darlington configuratons are shown in Fig. 4-20.

For switch-mode power supplies, it is another story—the analogy can lead to serious trouble. For example, even though spec sheets refer to the "saturation voltage," $V_{CE(sat)}$, of the power darlington, it is better to state that this device cannot be driven to saturation in the same sense as the discrete transistor. The output of the darlington comprises the E-B junction of the output stage in series with $V_{CE(sat)}$ of the driver stage. Another way of viewing this is to recognize that the collector-base junction of the output stage can never become forward biased because it is held to $V_{CE(sat)}$ of the driver stage. Thus, the use of the power darlington in a low-voltage switch-mode regulator can produce unacceptably high dissipative losses during on time, even if the switching losses are negligible at the frequency of operation. For higher-voltage operation, the dissipation because of the relatively high voltage drop across the darlington switch assumes less significance.

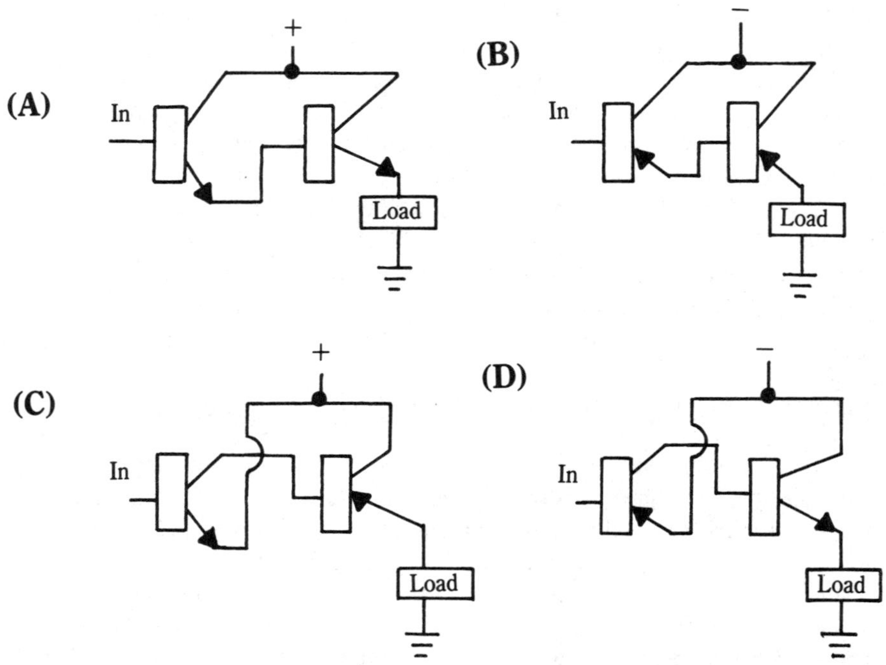

4-20 The basic darlington circuits.

Although the nonsaturation of the darlington output stage imposes a penalty of higher conductive loss, it enables higher switching rates to be attained than otherwise would be the case. To attain optimum switching performance, and particularly maximum switching speed, it is necessary to speed up removal of minority charges in the driver stage and in the base-emitter portion of the output stage. This can be done in three ways.

The best scheme for reducing storage time in the darlington transistor is to incorporate a monolithically formed network of two resistors and a diode as shown in Fig. 4-21. When a short-duration negative pulse is applied to the base terminal of the darlington, the minority charges in the driver stage are swept out. Because of the diode, D1, minority charges in the base-emitter portion of the output stage are also swept out. The overall effect is a shortening of the storage times that would otherwise seriously delay turn off of the darlington.

Another speed up technique does not require the internal diode-resistor network of Fig. 4-21. Instead, a high amplitude negative pulse is applied to the base terminal of the darlington so that the base-emitter section of the driver transistor is avalanched. This quickly depletes the stored charges in both the driver and the base-emitter section of the output transistor. There is some controversy over the wisdom of applying this technique because such cyclic avalanching has been associated with gradual failure of the driver transistor. Probably it is a good precaution to see to it that excessive drive is not applied beyond that needed for avalanching.

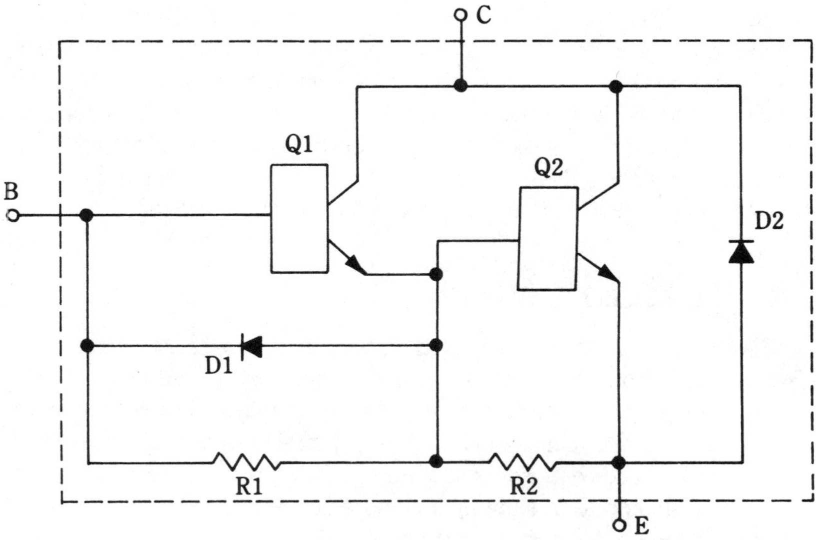

4-21 A power darlington together with the "bells and whistles."

A third method for reducing storage time involves bringing out the base of the output transistor to a fourth terminal to which a negative-going pulse can be applied. In some instances, the manufacturer will do this on request. Circuits then can be devised that will deliver negative pulses to the bases of both of the darlington transistors following completion of the desired on time.

Because darlingtons are particularly susceptible to thermal runaway, attention to adequate heat removal is a must. Whereas discrete power transistors often can be safely operated at a junction temperature of 200°C, a reasonable maximum junction temperature of darlingtons would be on the order of 150°C. Cooler junction temperature also decreases storage time of minority carriers, thereby enhancing switching performance.

All things considered, the power darlington is an electrically rugged device, providing that its unique needs are complied with. Most difficulties encountered stem from the misleading notion that it can be treated as if it were just a high-beta discrete transistor. Probably the most important consideration is to provide adequate heat removal. Power darlingtons in TO-3 packages carry power dissipation ratings up to 250 watts; 1000-volt, 50-ampere ratings are readily available. These devices are excellent for service in off-line converters operating in the 20-kHz to 40-kHz range. The easy-drive characteristic dispenses with expensive and technically difficult drive provisions. Indeed, operation directly from logic circuitry is feasible. Both npn and pnp types are available, and pairs can be used with sufficiently similar characteristics to give good performance in complementary-symmetry circuits. Reading between the lines, the opportunities for low cost and simplified design are quite evident.

Finally, another source of trouble from viewing the power darlington as the equivalent of a high-gain single transistor should be mentioned. The gain roll off beyond the "3 dB down" response is greater for the power darlington than it is for the discrete transistor. Because of this, the not uncommon practice of merely substituting the darlington for a proven single-transistor design can lead to instability and oscillation. Generally speaking, the feedback loop compensation will be different for the darlington than for the single transistor application. Another way of stating this is that the darlington tends to cause more phase shift at higher frequencies than does the single power transistor.

The BIMOS switch

The BIMOS switch, like the power darlington, is a two-device circuit. Unlike the power darlington, it has not been successfully fabricated as a monolithic structure. Despite the discrete nature of its topography, it is a very interesting implementation, and it fulfills a unique need identified by voltage, current, frequency capability, and cost. For example, when inverters, converters, or power supplies are called on to provide multiampere currents at a kilovolt, or so, and to respond to tens of kilohertz, a basic problem arises.

Discrete bipolar transistors can meet such requirements, but not cheaply. Such transistors border on the exotic and are inherently costly. Power MOSFETs certainly pose no speed problem, but tend to exhibit high values of R_D so that a voltage drop of 25 volts is not untypical. Because power dissipation increases as the square of the voltage drop, serious inroads into operating efficiency can result. The power darlington often can fill the bill, but the BIMOS siwtch probably has superior speed. Considering all factors in those applications where both of these dual devices merit consideration, selection might well be made according the relevant priorities—these could be efficiency, parts count, anticipated reliability, second sourcing, or overall costs.

The BIMOS configuration is premised on the cascode circuit, which has served special purposes since electron-tube technology. It involves a certain

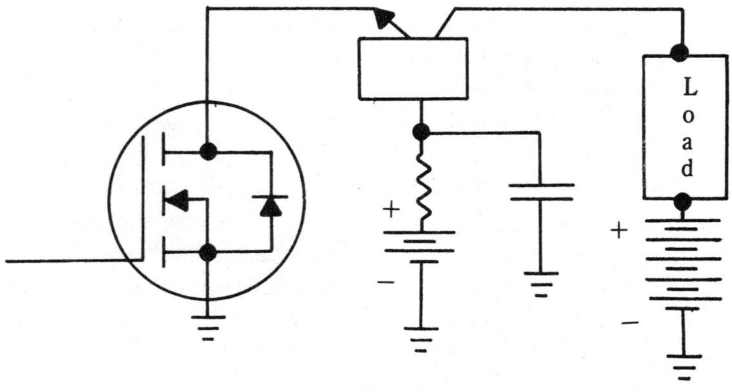

4-22 The basic BIMOS switching circuit.

way of directly coupling one active device to another. In the electron-tube version of the cascode circuit, the first, or input stage, was a "normal" grounded-cathode amplifier; the second, or output stage, was a grounded-grid amplifier. The analogous arrangement of the BIMOS switch uses a grounded-source power MOSFET as the input stage, and a grounded-base bipolar transistor as the output stage. See Fig. 4-22. What, indeed, is gained by such an unorthodox combination?

Consider, first, the nature of the bipolar output transistor. Most of us have known at one time that the bipolar transistor can withstand considerably more voltage, perhaps on the order of twice as much, in the grounded, or common-base circuit than in the grounded, or common-emitter circuit. Moreover, the frequency response of the grounded-base circuit is much better than in the grounded-emitter circuit. The drawback, however, is that it is difficult to drive the low-impedance input of the grounded-base circuit; this is particularly so when such an amplifier circuit must provide appreciable power to the load.

At first thought, a transformer might appear as a means of stepping up the inordinately low input impedance of the grounded-base stage to a more manageable level. This has some merit, but in practice it is easier said than done; new troubles arise from leakage inductance, resonances, limited bandwidth, and copper and core losses. A better approach is the use of an "electronic transformer" in the form of a low-voltage power MOSFET. When the two active devices are combined in the cascode configuration, you have the best features of each. The bipolar transistor can be a relatively inexpensive type because the grounded-base connection is much less demanding of withstanding voltage than is the more conventional grounded-emitter circuit. On the other hand, the power MOSFET can be a relatively low voltage type; its use imparts a very high impedance input to the overall circuit—quite a different situation than driving a grounded-base stage directly.

Note, too, that the grounded-base bipolar transistor produces a small fraction of the voltage drop that would be associated with a high-voltage power MOSFET. It is true that the overall voltage drop of the composite circuit is actually the sum of the voltage drops developed in the MOSFET and bipolar transistor, but this is still small compared to the drop that would be incurred in a high-voltage MOSFET.

Another nice feature of the duo is that the overall frequency response is about what could be expected from just the grounded-base bipolar transistor alone. All things considered, you have an inexpensive and efficient switch for the 1000-volt region with good speed capability. This basic circuit has been used for the horizontal deflection driver in TV receivers, where it promises to alleviate a long-enduring reliability problem in this application.

The voltage-biasing source for the base of the bipolar transistor is employed to overcome the turn-off bias produced by the voltage drop across the drain-source terminals of the power MOSFET. In other words, R_D of the MOSFET acts as a biasing resistor in the emitter circuit of the bipolar transistor.

Faithful reproduction from transformers

Power-supply designers sometimes get by with minimal attention directed to transformer applications; often a high-frequency version of a simple bell transformer does the job. On the other hand, the most sophisticated approach to transformer design can result in "mysterious" behavior of switching power supplies, often accompanied by failure of the switching transistors. This is not uncommon even after inordinate attention has been given to transformation ratio, winding impedances, distributed capacitance, linearity, saturation, and leakage inductance. What, indeed, has been overlooked?

In the evolution of transformer theory, most of the practical concern was with sine waves, and somewhat with square waves when audio technology became important. Square-wave response was useful because it gave an inkling of the high-frequency capability of the transformer. However, with the advent of pulse-width modulated power supplies, a new consideration became important. This had to do more with the application than the basic design of the transformer, but pertains to a characteristic of transformers not often dealt with in other equipment. This characteristic is somewhat subtle in the sense that it is not likely to be easily recognized from an oscilloscopic display of primary and secondary waveforms.

Suppose that an adjustable duty-cycle pulse generator is connected to the primary of a transformer and the primary and secondary voltages seen on a scope are observed as the duty cycle or pulse width is changed. See Fig. 4-23. Providing the amplitude of the pulses from the generator are kept constant, the voltage monitored across the secondary winding also will maintain a fixed peak-to-peak amplitude. Careful evaluation, however, will reveal that the relative amplitude of the positive and negative excursions change with the duty cycle. The transformer does this to preserve equal volt/second areas in the positive and negative portions of the waveform. This is as innately a transformer characteristic as is transformation ratio, but it is suprising how often it is overlooked. Obviously, if it is desired to apply 10 volts positive to the gate of a power MOSFET switching transistor through a transformer alone, the actual gate voltage can be "all over the map" during pulse-width modulation. Here lies the "mystery" of power supply malperformance and device failure.

Duty cycle independent drives for MOSFET power switches

The simple circuit shown in Fig. 4-24 provides one means of overcoming the secondary voltage difficulties when the driver is transformer coupled to the power switch in PWM supplies. Here, Q1 is a small MOSFET and Q2 is the power switch. The body diode of Q1 and the input capacitance of Q2 are shown in dashed lines inasmuch as these parasitic parameters play an important role in operation. Another circuit feature is the "inadequate" transformer; this transformer is purposely made very small, with insufficient core volume to support the

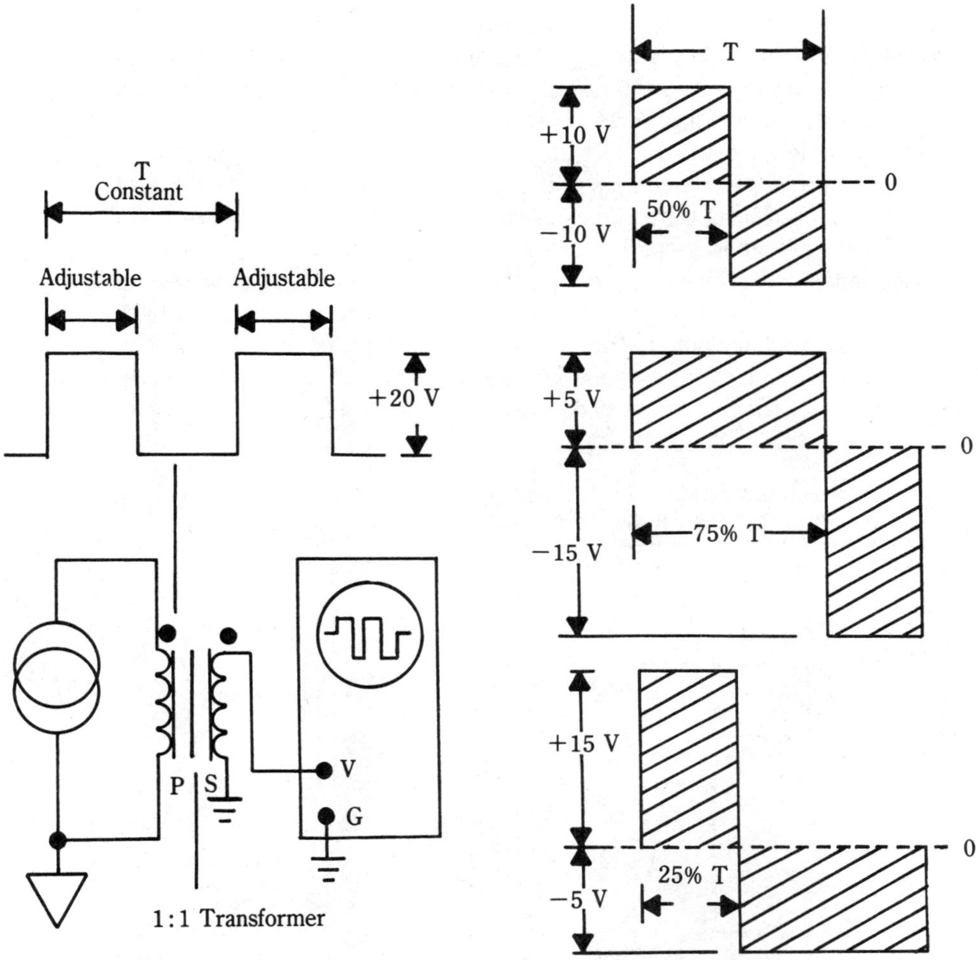

4-23 Nonfaithful response of a driver transformer to a PWM wave.

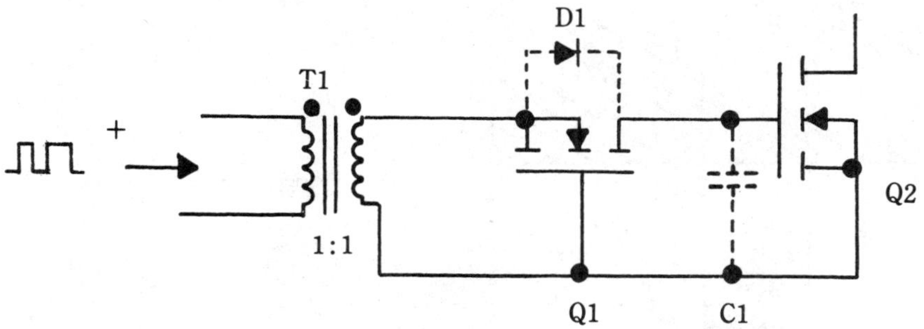

4-24 Wide-range PWM drive circuit for power MOSFET switch.

duration of the narrowest pulse to be encountered in the control process. This helps keep this circuit application compact and economical. It is capable of providing nearly constant enhancement-voltage pulses to the gate of the power switch throughout a duty cycle range of well over 90%.

The equivalent circuit of Fig. 4-25 leads to ready insight into the unique operating principles. Here, the physical switches represent the body diode of Q1 and D1, itself. The closure of these switches takes place alternately—when one is closed, the other is open. Furthermore, the upper switch representing the body diode, D1, closes only during the positive excursions of waveform "B" (because of the saturation characteristics of the undersized transformer, wave "B" is a partially differentiated version of the primary wave, "A"). The lower switch, which represents the transistor portion of Q1, closes only during the negative excursion of the secondary voltage-wave, "B."

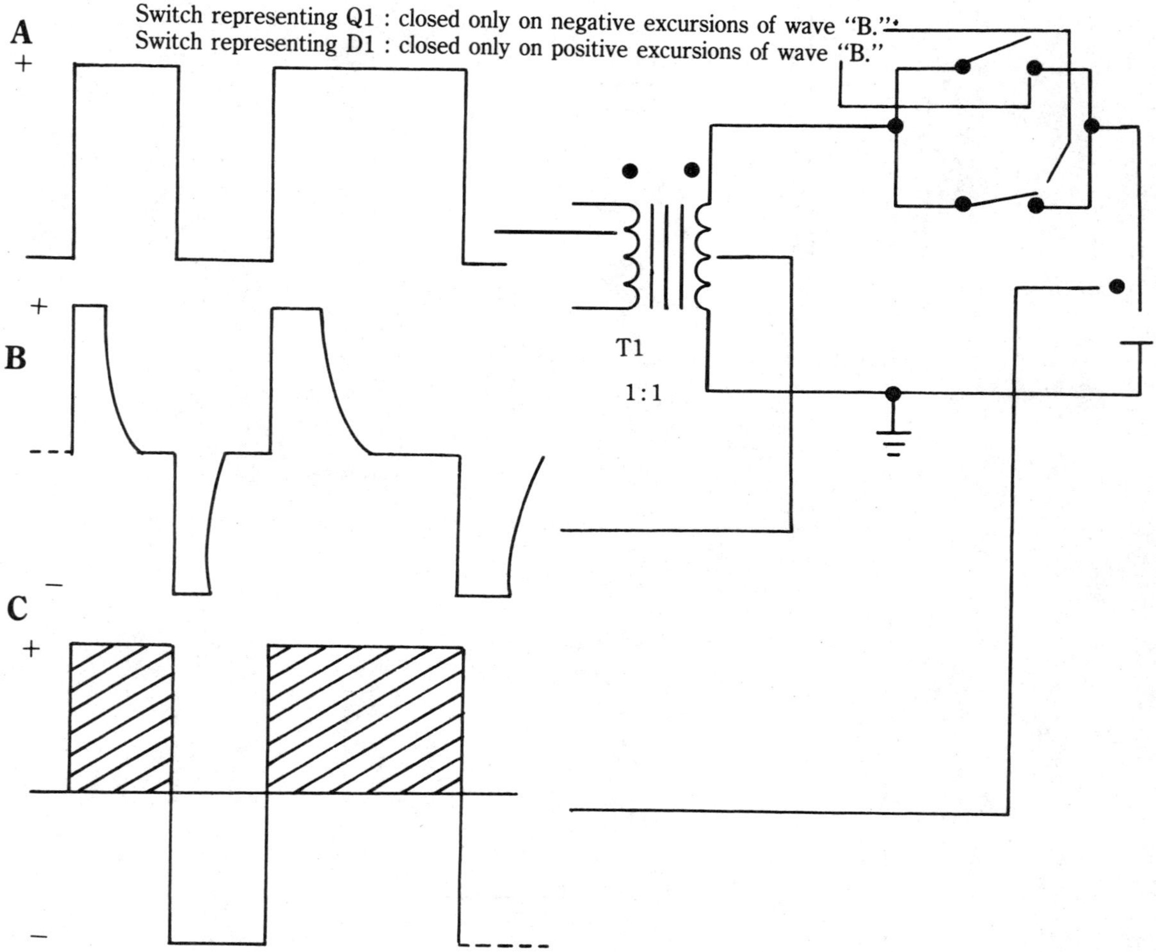

4-25 Equivalent circuit of wide-range PWM drive circuit.

The synchronization of the alluded switching actions are brought about by polarity changes "seen" by the body diode, D1, and the gate of Q1. When the body diode permits conduction, capacitor C1, which represents the input capacitance of the power switch, is quickly charged to the positive value needed for strong enhancement of the power switch, Q2. Although this charging process is the result of a very narrow pulse, C1 "remembers" the voltage level. The "remembered" voltage level is abruptly terminated by the arrival of a negative pulse when the lower switch (Q1) closes. In this way, the positive excursions of the waveform developed across C1 faithfully follow the timing sequence of the PWM wave, "A." The positive and negative portions of wave "C" at the capacitor remain constant in amplitude with respect to the width of the PWM pulses.

Proper operation of this circuit requires a low-impedance source for the PWM wave. This is especially true for high switching rates. The totem pole outputs from logic circuits are well suited for this purpose.

Another scheme for securing the isolation feature of a driver transformer without incurring the penalty of varying drive voltage in PWM systems is depicted in Fig. 4-26. The operating principle is altogether different from the circuit of Fig. 4-24. In Fig. 4-26, something is done to the primary voltage, rather than to the secondary voltage as in Fig. 4-24. Although shown as a drive technique for a pair of power switches, the circuit is also workable for single-

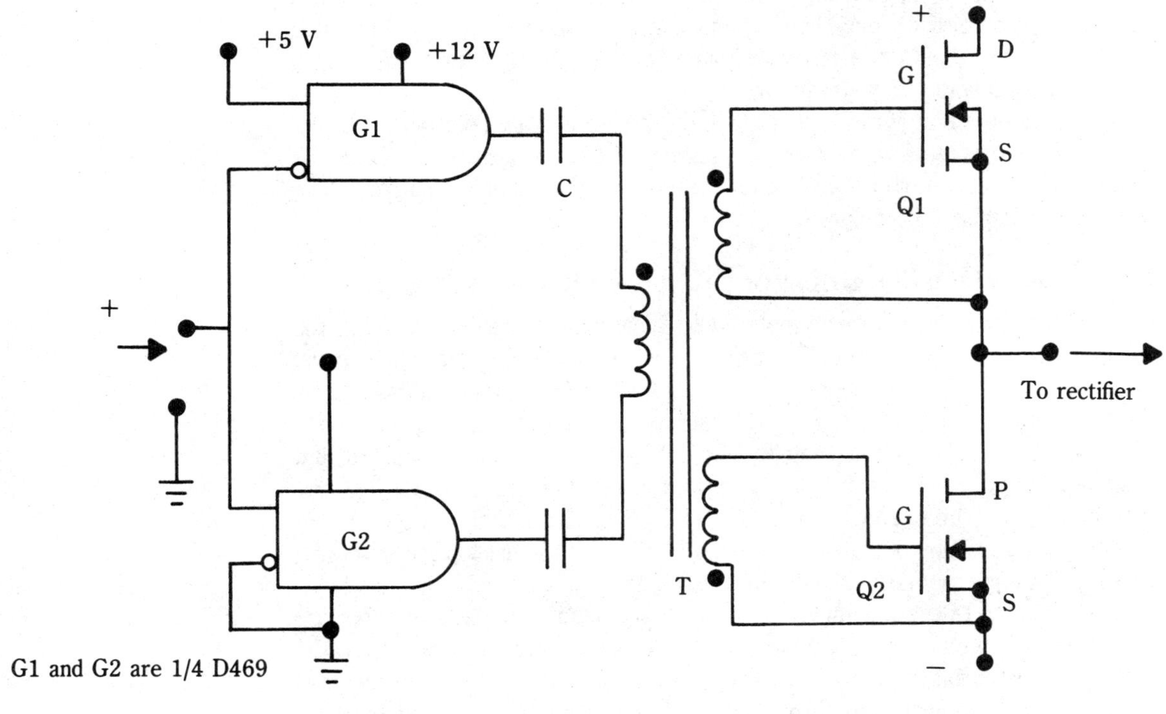

4-26 Another constant-amplitude driver for PWM supplies.

ended output. This could be achieved by dispensing with the lower secondary winding, Q2, and the negative supply.

Essentially, the logic gates convert the unipolar PWM wave to bipolar form. This makes the transformer "happy" in the sense that there is no DC component to affect core reset. The bipolar wave is a true AC wave, allowing equal watt-second excursions for both halves of the cycle. This being the case, the transformer is designed to provide faithful reproduction of the bipolar PWM wave. The two coupling capacitors are shown to emphasize the basic symmetry of the circuit; actually a single capacitor will suffice. It should be large enough to preserve the integrity of the PWM wave.

The logic gates require dissimilar polarities to change state. Therefore, the unipolar PWM wave will cause the gates to alternate their output logic levels—when the top gate is high, the lower gate will be low, and vice versa. There is no danger of overlap or simultaneous conduction. The primary of the transformer interprets this action as an impressed AC wave with symmetrical positive and negative excursions. If the two remaining gates in the quad are not used for other purposes, they can be paralleled with the gates used in this circuit. This would provide a lower output impedance and speeds up switching of the power MOSFETs.

Note that this circuit cannot work unless the transformer windings are phased as indicated. Generally, a one-to-one transformation ratio suffices. (A slight step-up ratio can be advantageous in driving the power switches harder to lower their R_D.) Close coupling between the windings is essential, and it is a good idea to empirically establish minimal turns needed to faithfully reproduce the PWM wave. Sometimes resistive loading of the secondaries can prove beneficial in suppressing ringing effects.

Although the circuits of Figs. 4-24 and 4-26 are elegant solutions to transformer drive problems when a wide-ratio control of duty cycle is involved, the fact remains that a simple transformer alone can be used if only a narrow range of duty cycle change is required.

The "mysterious" demise of three-terminal regulators

Throughout the years, power supply books have dealt rather heavily with the intricacies of linear power supplies, primarily with the familiar topography using a series-pass element as an electronically controlled "rheostat." At first, these supplies were entirely designed with discrete transistors; later op amps and other ICs were used to both simplify and improve these closely regulated, but relatively inefficient supplies. With the advent of the three-terminal voltage regulator, there has been considerable availability of technical literature on linear power supplies. This, of course, has been because of the availability, low cost, and ease of implementation of the three-terminal regulator.

With regard to ease of implementation, these handy monolithic devices are all too often "thrown in" to a system, and often as an afterthought. Thus, the voltage regulator shown in Fig. 4-27a represents the extreme in simplicity—it is merely interposed in the circuit with the understanding that the input (unregulated) voltage is higher than the desired output (regulated) voltage. Such a primitive approach often works satisfactorily. However there is a good chance also of unstable behavior. This is often the consequence of a long input lead (several inches will suffice) and/or the lack of an output capacitor. An output capacitor should not be thought of as a filter capacitor, but rather as a means of pacifying the phase-gain relationship in the feedback loop.

The addition of these capacitors is shown in Fig. 4-27b. Generally, a 0.1-μF ceramic capacitor suffices for the input capacitor, C1. 1.0-μF solid-tantalum capacitors also work well in this part of the circuitry. This prescription tends to be suitable for the output capacitor, C2, as well. However, some three-terminal regulators require 10-μF solid-tantalum output capacitors to operate stably under varying load and input voltage conditions. Aluminum electrolytic capacitors are less desirable; if they are used, about 25 times more capacitance is required for similar results forthcoming from solid tantalums. Also, it is a good idea to shunt aluminum electrolytics with 0.1-μF ceramic capacitors to maintain high-frequency bypass action.

A side effect of Fig. 4-27b is that the regulator is now vulnerable to catastrophic destruction under certain fault conditions. If the input to the regulator

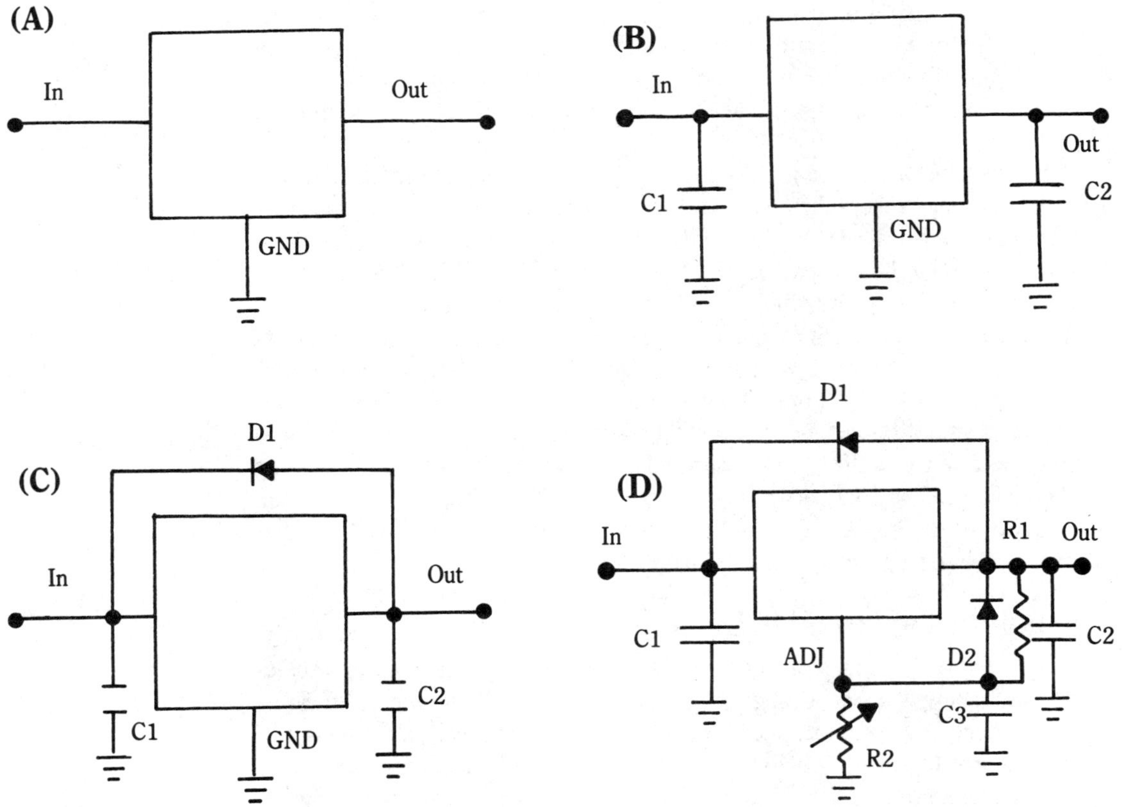

4-27 Applications of three-terminal voltage regulators.

should be inadvertently shorted to ground, output capacitor, C2 then will discharge through the regulator with an often-damaging peak current. Similar destruction or damage could occur if the output voltage should become higher than the input voltage.

Although operational stability might prevail without C2, it usually is not desirable to omit it just as a safeguard against discharge damage to the regulator. This is because the presence of this output capacitor usually is manifested by greatly improved transient response. Yet another way the regulator in circuit 4-27b can suffer damage is by plugging it into a ''live'' socket; if the GND lead happens to be the last to make contact, the output voltage will quickly approach the unregulated voltage at the input. Damage then can occur when the GND lead makes contact. Note: if the GND lead is first connected, it is then allowable to connect the other two terminals in any sequence and at any time.

A simple and inexpensive protection against any of the damage-prone conditions described here is shown in Fig. 4-27c. The diode can be a small general-purpose type such as the 1N4002. In the case of a negative-voltage regulator, the diode polarity would be reversed from that shown. During ordinary operation, the diode has no effect on the performance of the regulator.

The regulator depicted in Fig. 4-27d is an adjustable type. Here, capacitor C3 confers a worthwhile reduction in output ripple. Protective diode D1 functions as in the previous circuit. In this arrangement, C3 can deliver a destructive discharge current if either the input or output of the regulator is shorted and D2 is not present. Typical values for R1 and R2 are 240 ohms and 5 k respectively.

The previous discussion is presented as an explanation to why three-terminal regulators sometimes die a sudden and mysterious death. It is evident that despite their electrical ruggedness, and their built-in overload and temperature limiting, they nonetheless have their Achilles' heel. Protective diodes always should be used during breadboard tests and during experimentation and, in most instances, in the final product as well. An interesting use of the three-terminal regulator is for postregulation of the output of a switch-mode power supply. Such a supply can still operate at a reasonably high overall efficiency, but can deliver a smoother output than would be attainable from the switcher alone. In a very real sense, a three-terminal regulator used in this way functions as an electronic filter. Being outside of the feedback loop, it introduces little effect on the stability of the supply. Special three-terminal regulators are available that minimally degrade efficiency when utilized in this fashion.

A look into the crystal ball

It is easy enough to indulge in prognostication of future developments in electronics technology—devices and systems obviously will be more efficient, speedier, more powerful, and more reliable than their ancestors. Forecasts of this nature are, however, nebulous; although they can't be readily refuted, it is not easy to relate them to the practical hardware of today. An aircraft designer, for example, would experience greater relevance to his pursuits from reading about research in composite materials as metal substitutes, than acquainting himself with the latest speculations about antigravity machines.

This being the case, let's look at possible *and* probable developments in power-supply technology. This will be accomplished by limiting our predictions to extrapolated developments of devices and techniques already in existence.

1. The immediate future will see continuing momentum in development of high switching rate supplies. The 100-kHz to several-MHz region will be explored for optimum balance of performance, compactness, reliability, and cost. Different applications will favor different frequencies, but it appears likely that 10 MHz or so will represent an area of diminishing returns. On the other hand, several hundred kHz appears to be an operating region that readily accommodates circuit components, is not too demanding on production methods, and can allow about the same power levels as lower-frequency operation. Thus, in place of the long-popular 20-kHz to 40-kHz switchers, a tenfold increase in switching rate can be anticipated.
2. Resonant-mode topography will serious compete with the long-enduring PWM switching technique. More dedicated-control ICs will become available for resonant-mode operation as awareness of its advantages at high frequencies spreads.
3. It has been determined that computers can operate at higher speeds and with less internal dissipation if logic circuitry is designed for 3-volt, rather than 5-volt operation. This will create a demand for 3-volt power supplies and more severe demands will be imposed on power switches and on rectifiers. These will have to produce very low voltage drops so as not to impair the operating efficiency of such low-voltage supplies. More emphasis on synchronous rectification will bring forth more power MOSFETs optimized for this purpose. At the same time, linear supplies and germanium transistors will experience renewed consideration.
4. Thermal techniques will assume increased importance as power densities increase because of higher operating frequencies. Copper can be expected to replace aluminum for heatsinks in many instances where it is not feasible to penny-pinch costs. Diamond, which possesses inordinately high thermal conductivity might find its way into spacecraft and military power supplies. Also, more use will be made of heat pipes and of thermoelectric coolers. For certain applications, refrigeration will allow enhanced power densities at high operating efficiency including the refrigeration process. Liquid nitrogen temperature appears to be the easily achievable, and the most economic, refrigeration to implement.

   At the other extreme, it has been demonstrated that gallium-arsenide devices can operate at higher temperatures than silicon devices. Manufacturing difficulties have slowed development of gallium-arsonide counterparts to silicon power devices, but a breakthrough in fabrication methods conceivably could make such devices a common-market commodity.

5. Power op amps and "smart" power devices are already with us, and are in the process of "growing up." Although worthwhile progress in simplicity and production economics accrues from placing much of the power supply on a single chip, it is both an art and a science to overcome the thermal and electrical barriers that quickly assert themselves with attempts to design for higher power levels. Dramatic accomplishments already have been made in this field; examples are the power darlington (actually a monolithic IC), the three-terminal voltage regulator, "synthesized" transistors (such as the LM195), and a number of op amps intended for operation of motors.

   It is also likely that light-controlled power devices will receive emphasis because of the electrical isolation they provide between control logic and load circuitry. These, too, already exist. For example, utilities use banks of optically actuated SCRs for switching functions involving very high currents and voltages. Because of semiconductor lasers and fiberoptics, optically actuated devices are more practical than they were before. The optoelectronic isolators often used in feedback loops of power supplies exemplify this technology, although their power capability is low. A scaled-up version would use a power transistor, rather than a signal-level device in the optically controlled output stage.
6. Expanded use of uninterrupted power supply (UPS) equipment is likely because of the crucial nature of certain loads such as hospitals, computers, military facilities, and so on. A UPS monitors the presence of power from the normal AC line; in the advent of AC power failure, or of abnormal AC voltage, automatic switchover to a standby AC power source occurs. The standby source can be an engine-driven alternator or generator with inverter, a battery with inverter, or an alternate AC power line. UPS equipment takes many forms—single and polyphase supplies are available. Circuit technology can make use of thyristors or banks of paralleled transistors. Sophistications include the sensing of phase so as to provide glitchless transitions, and line conditioning to ensure transient-free power from both the normal AC line, and the emergency source. In any event, emphasis is on fast switchover to spare critical loads any disruptive disturbance.
7. Inverters for electric vehicles using AC motors will become more competitive with DC motor technology. Such designs offer relatively maintenance-free operation and can be very efficient. Much, of course, depends upon improvements in the energy-to-weight ratio of batteries, to say nothing of the politics of petroleum supply. The use of choppers as the control means of DC motors also has much potential in traction vehicles, where giant germanium transistors can be used.
8. Nonmechanical utility stations using fuel cells and electronic inverters are already feasible. First applications probably will be for enhancing peak power capability of existent fossil-fuel generators. If properly

implemented, such technology can produce environmentally clean electric power on a low-cost, low-maintenance basis.

9. Because of ever-increasing RFI and EMI ''pollution,'' sine wave circuit techniques will challenge traditional square wave designs despite lower operating efficiencies.
10. At the other end of the power spectrum, demand will increase for regulators and inverters capable of providing efficient operation from a very low voltage DC source, say, one to three volts. Dedicated ICs for this purpose are already marketed. Their salient features are low quiescent current and low voltage drop. Both linear and switching topographies are useful. Load currents tend to be in the milliampere or microampere region. Such miniaturized power-supply technology will find application in solar-cell circuits, display units, bias sources, and in portable-computer designs.

These extrapolations of the present state of the art deliberately have been conservative. In a more imaginative vein, the unanticipated technological breakthrough remains just around the corner—there is nothing static about a ''static inverter!''

Appendix A
Schematic symbols

AMPLIFIERS

Operational

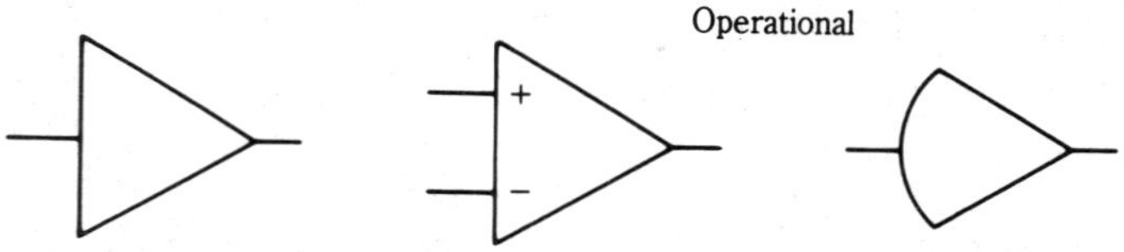

Single-ended
or buffer

(preferred)

ANTENNAS

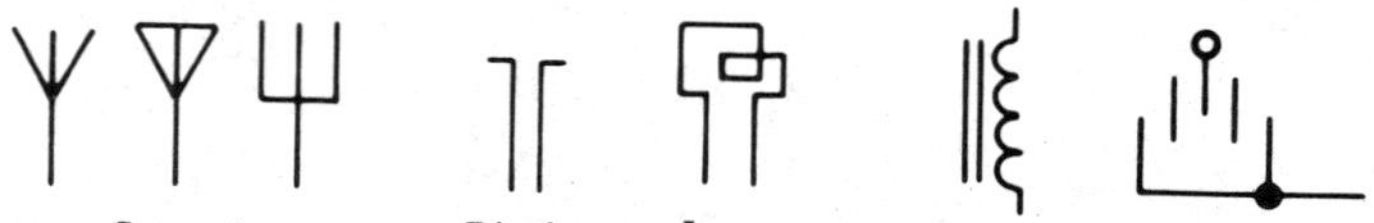

General Dipole Loop Ferrite Whip

BATTERIES

Single cell Multicell

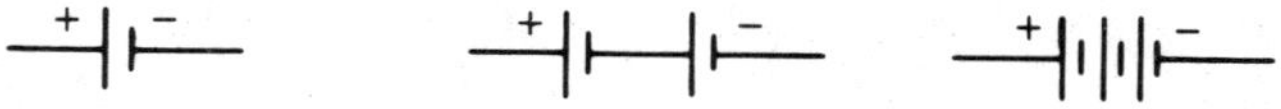

(preferred)

BELL

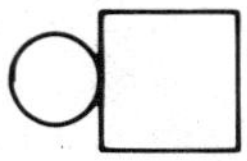

BUZZER

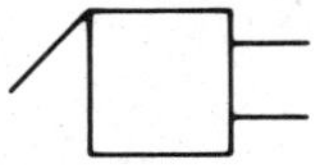

CABLE

Two-conductor
cable with grounded shield

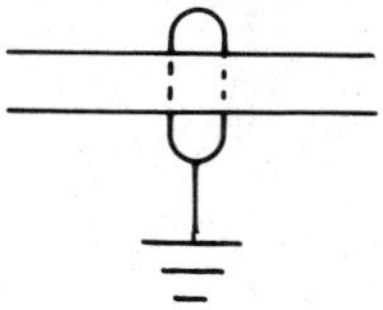

Coaxial cable
with grounded shield

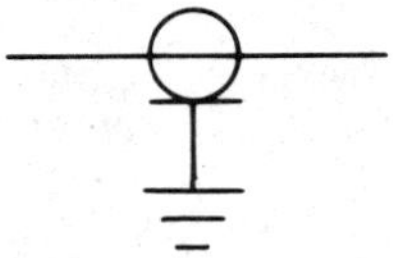

Twisted pair

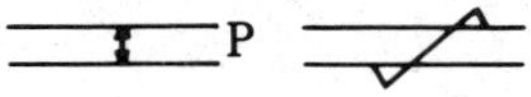

Stripline

(unbalanced)

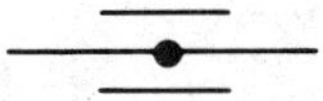

(balanced)

CAPACITORS

Fixed

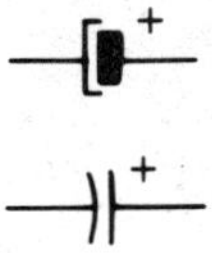

Electrolytic

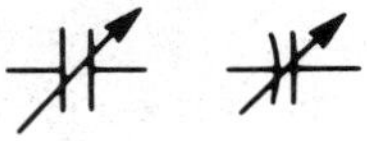

Variable

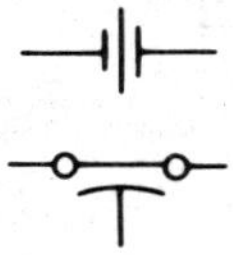

Feed through

Split
stator

Split
rotor

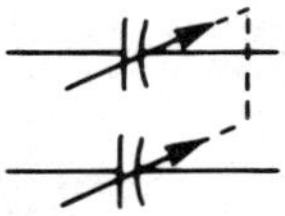

Ganged
variable

CIRCUIT PROTECTORS

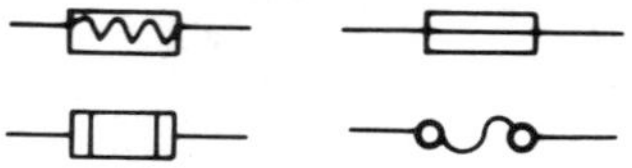

Fuse

Circuit breaker

CONNECTORS

Female contact

Male contact

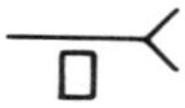

Jack or receptacle (usually stationary)

Plug (usually movable)

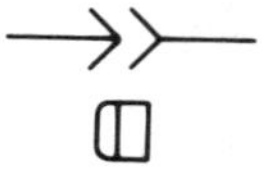

Separable connectors (engaged)

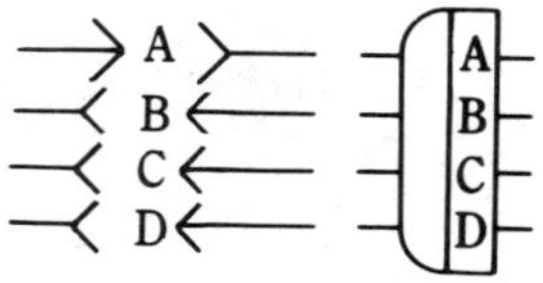

Four-conductor connectors (engaged) with individual contact designations

Common connection (ground)

Common connection

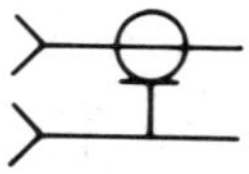

Coaxial receptacle

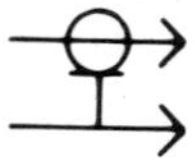

Coaxial plug

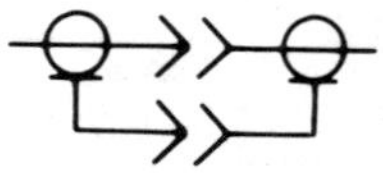

Coaxial connection (engaged)

Nonpolarized receptacles (117 V)

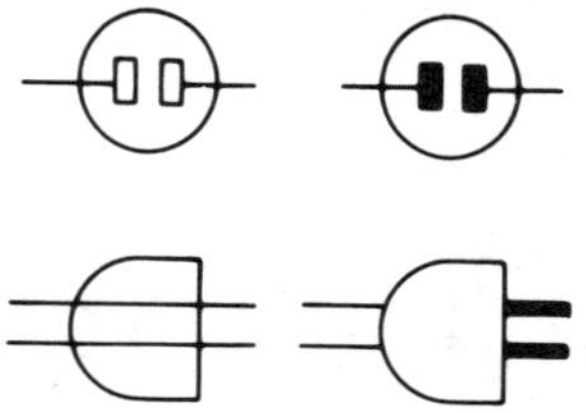

(Female) (Male)

Polarized receptacles (117 V)

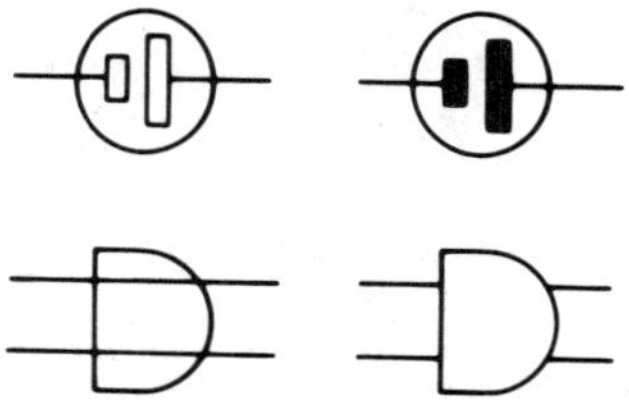

(Female) (Male)

Interconnecting plugs

(Female) (Male)

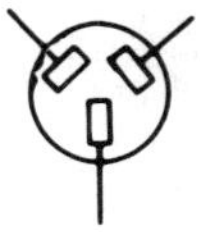

230 V plug

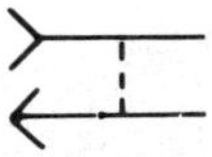

Phono jack

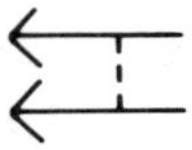

Mike jack

Phone jacks

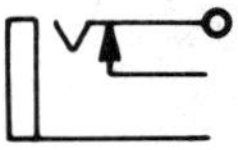

(closed circuit)

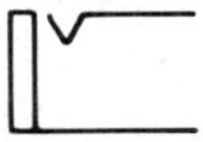

(open circuit)

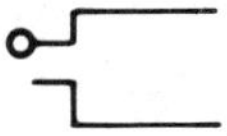

Phone plug

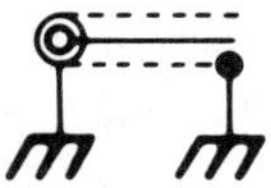

Phono plug

CONTACTS

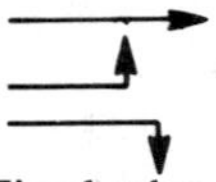

Fixed relay

Fixed switch

Moving contacts

(locking)

(nonlocking)

Closed contact

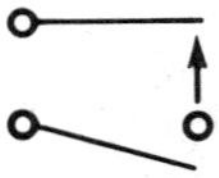

Open contact

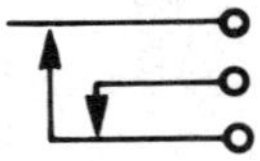

Make-before-break assembly

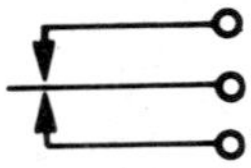

Break-before-make assembly

CRYSTALS

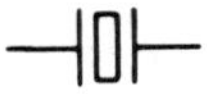

Piezoelectric

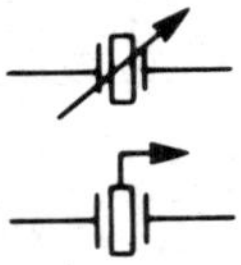

Monaural piezoelectric cartridges

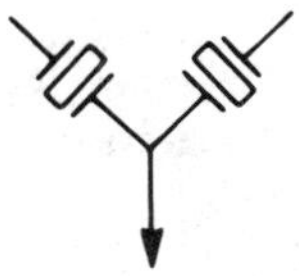

Stereo piezoelectric cartridge

DIODES

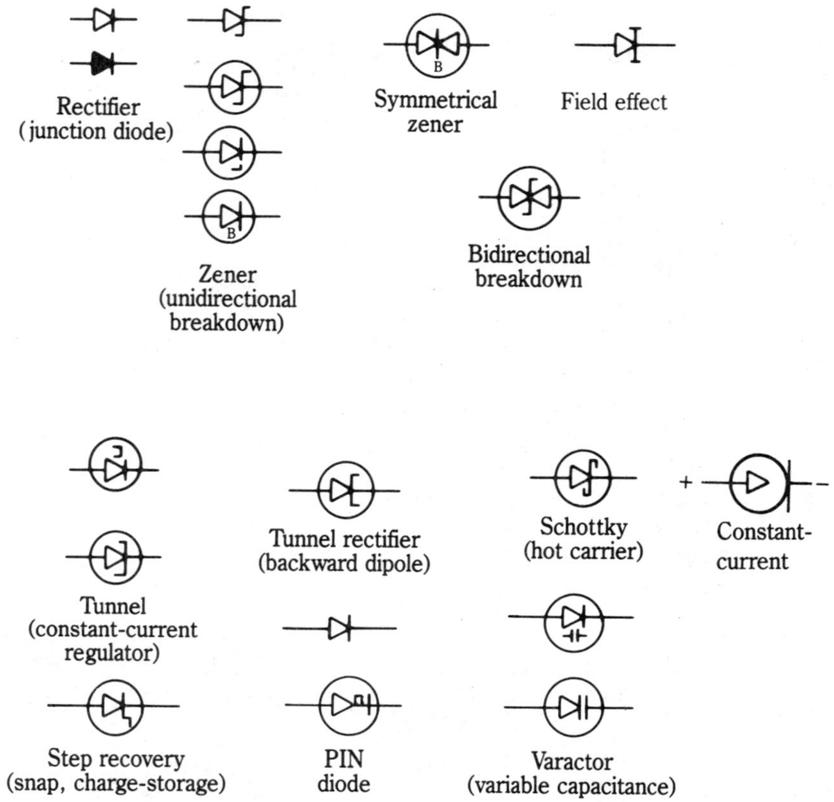

DIODES—OPTOELECTRONIC

Optically Coupled Isolators

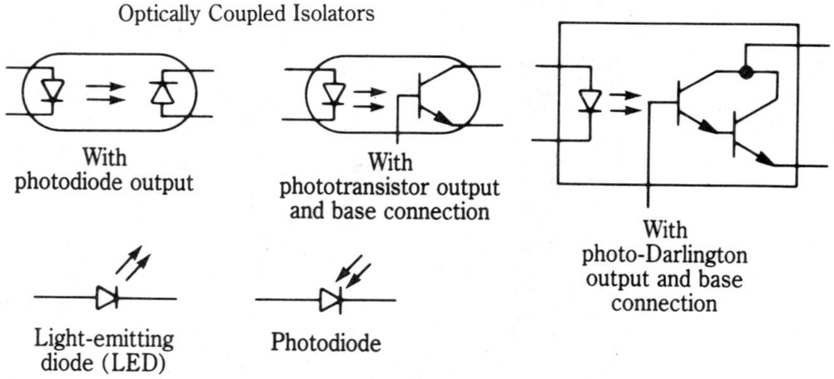

GROUNDS

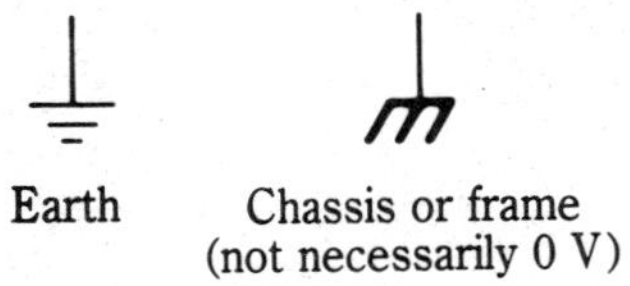

HEADPHONES

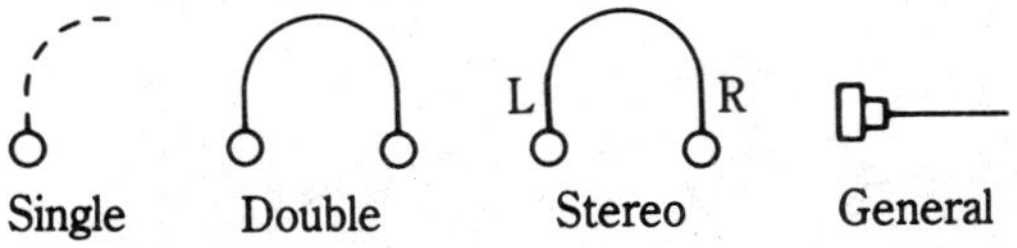

INDUCTORS (COILS)

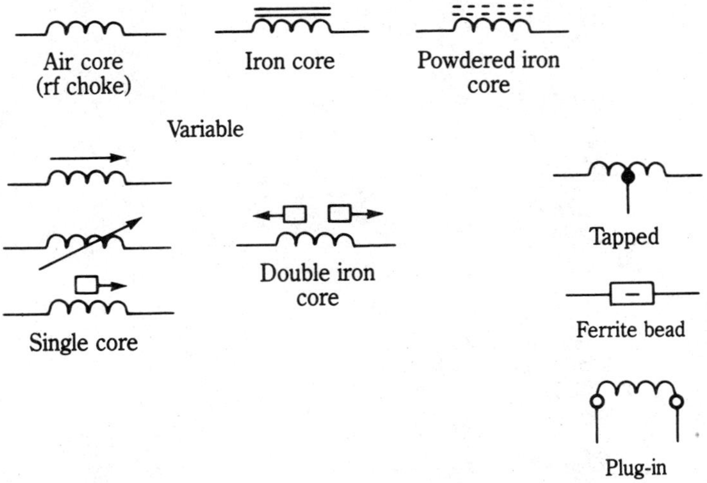

KEY

LAMPS

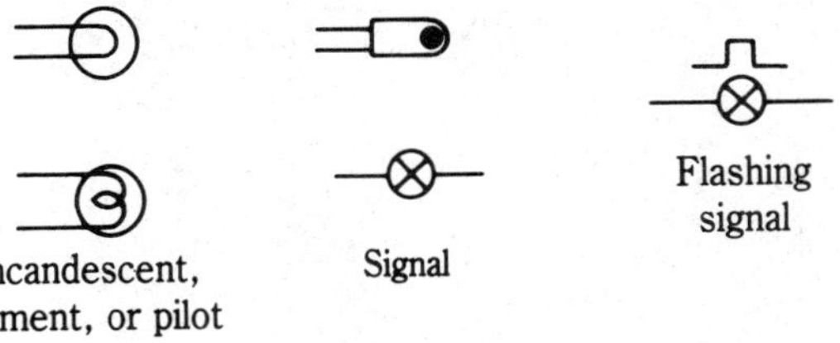

Neon

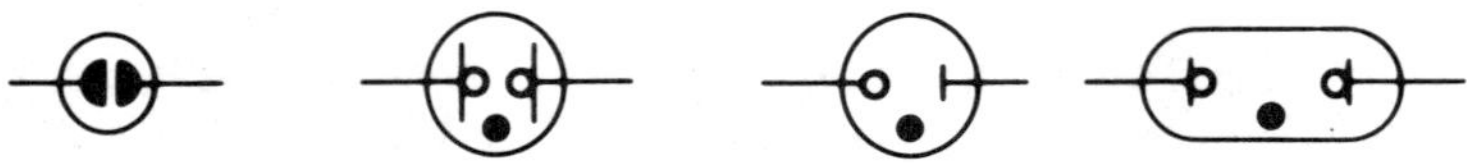

General

ac

dc

Fluorescent, two-terminal

LOGIC DEVICES

Gates

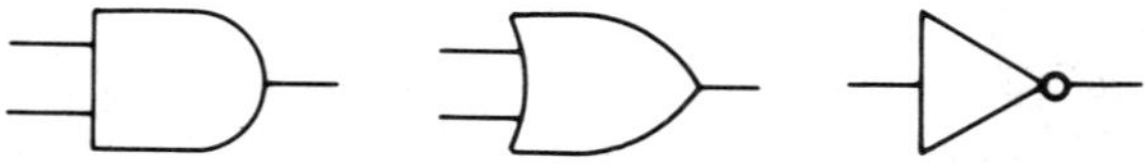

(preferred) (preferred) (preferred)

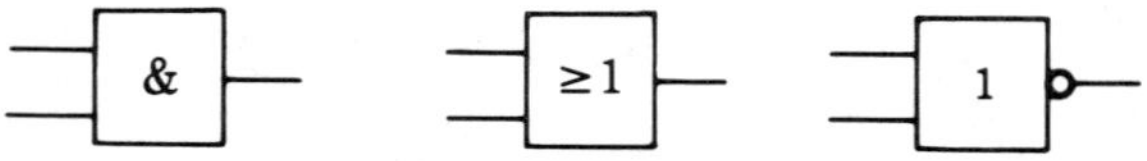

AND

OR

Inverter

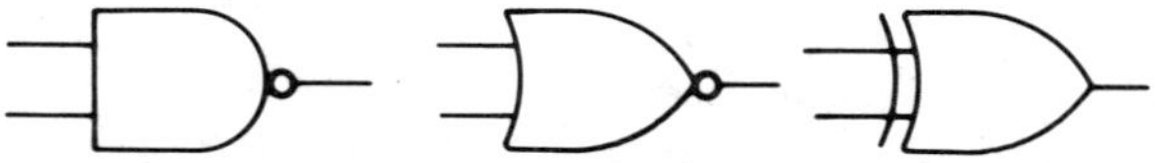

(preferred) (preferred) (preferred)

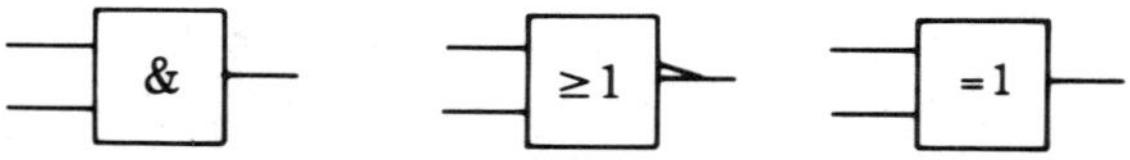

NAND, negated output

NOR, negated output

Exclusive OR

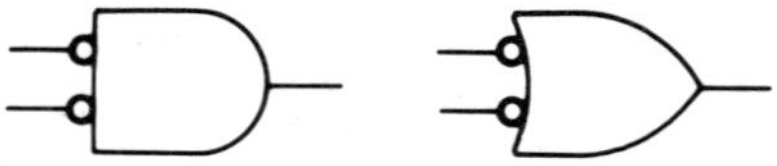

(preferred) (preferred)

Exclusive NOR

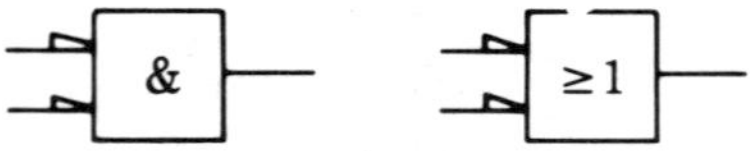

NAND, negated inputs

NOR, negated inputs

Flip-flops

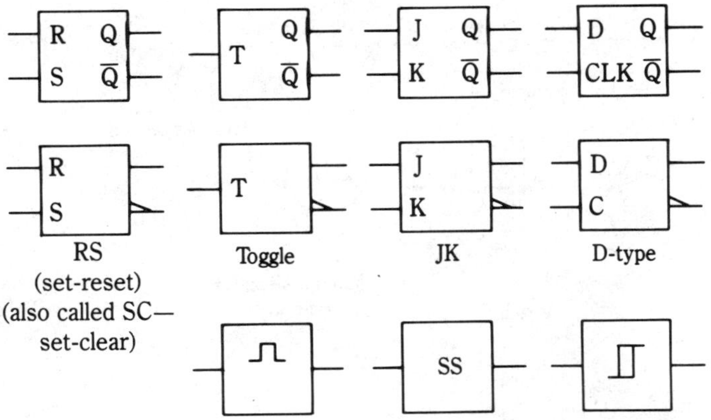

Monostable multivibrator (single shot, one shot)

Schmitt trigger

METERS

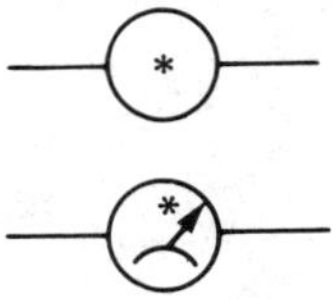

*Designations:

A—ammeter
mA—milliammeter
V—voltmeter
dB—decibel meter
etc.

MICROPHONES

General

Directional

MOTORS

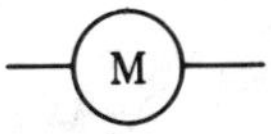

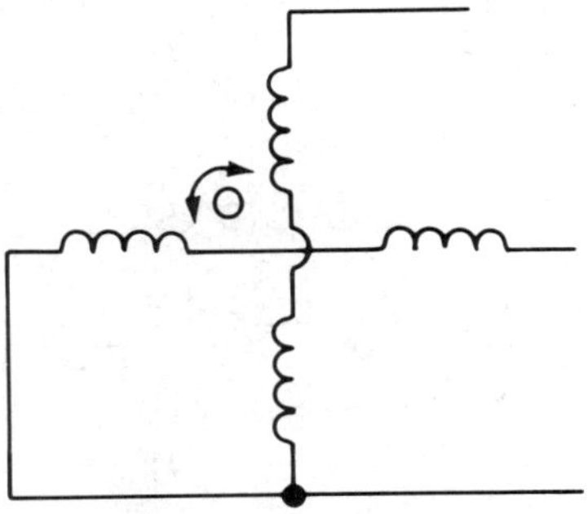

Reversible

RESISTORS

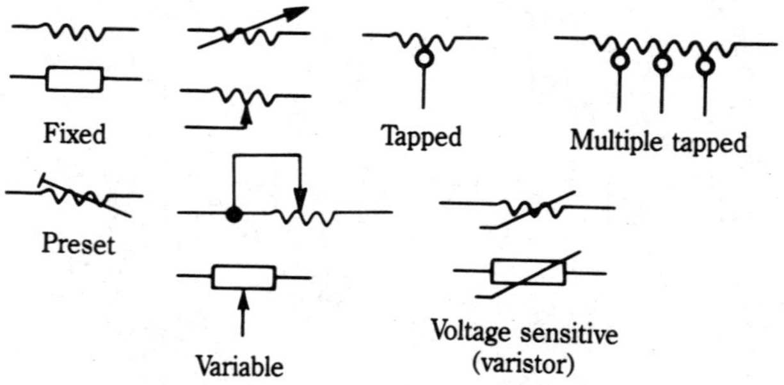

SPEAKERS

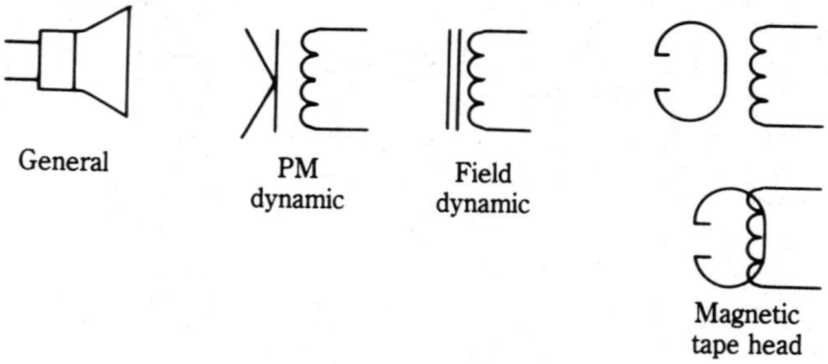

SOURCES

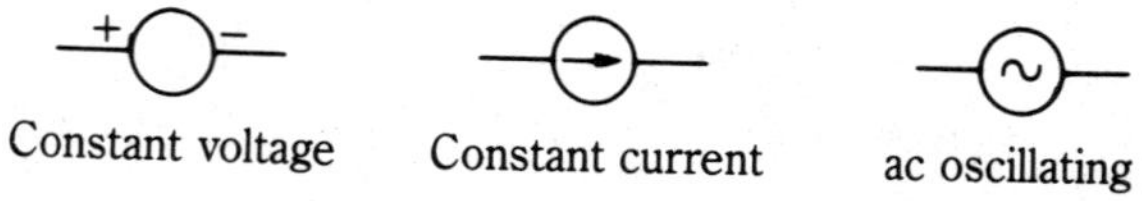

SWITCHES AND RELAYS

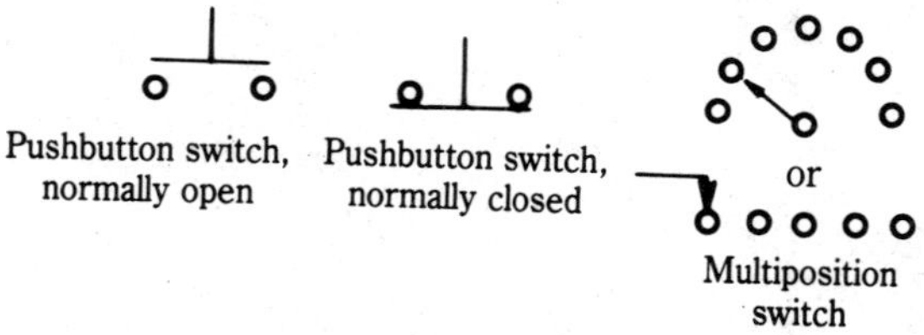

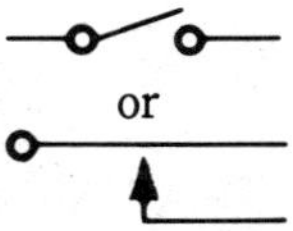

Single-pole, single-throw switch (SPST)

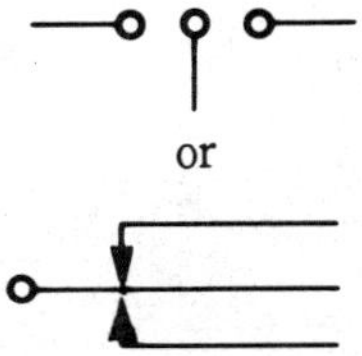

Single-pole, double-throw switch (SPDT)

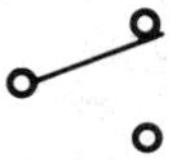

Single-pole, double-throw toggle

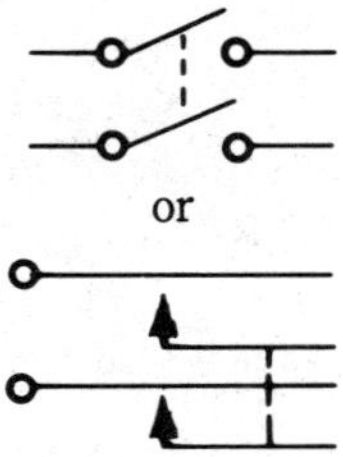

Double-pole, single-throw switch (DPST)

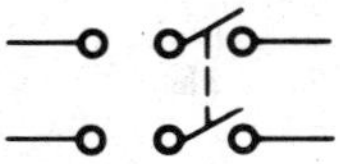

Double-pole, double-throw switch (DPDT)

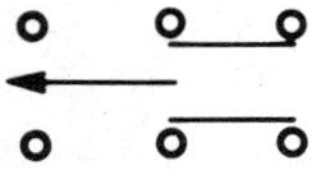

DPDT slide switch

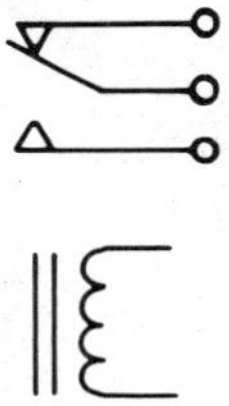

Relay (general)

Relay Contact Arrangements
(Heavy arrow indicates direction of operation)

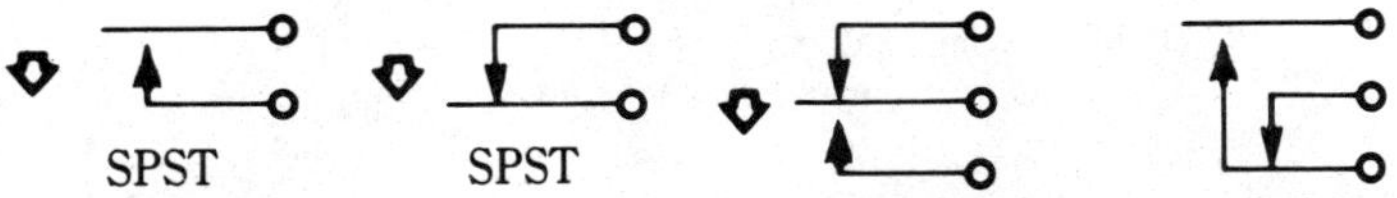

SPST normally open (make) | SPST normally closed (break) | SPDT (break, make) | SPDT (make, break)

TERMINALS (Board or strip; four terminals shown)

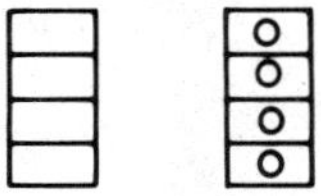

TEST BLOCK (Terminals shown)

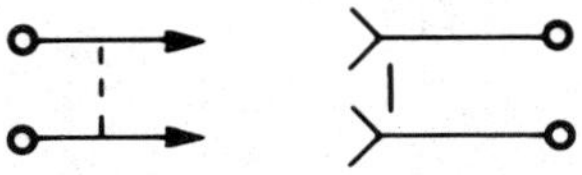

Male | Female short-circuiting bar

THYRISTORS

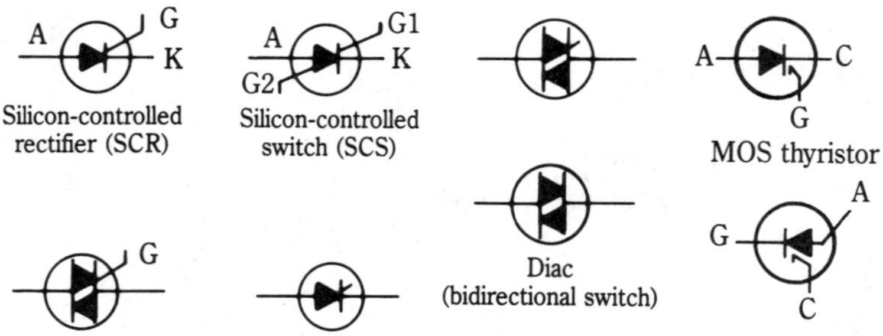

Silicon-controlled rectifier (SCR)

Silicon-controlled switch (SCS)

MOS thyristor

Diac (bidirectional switch)

Triac (gated bidirectional switch)

Four-layer diode (PNPN or Shockley)

MOS-controlled thyristor (MCT)

TRANSFORMERS

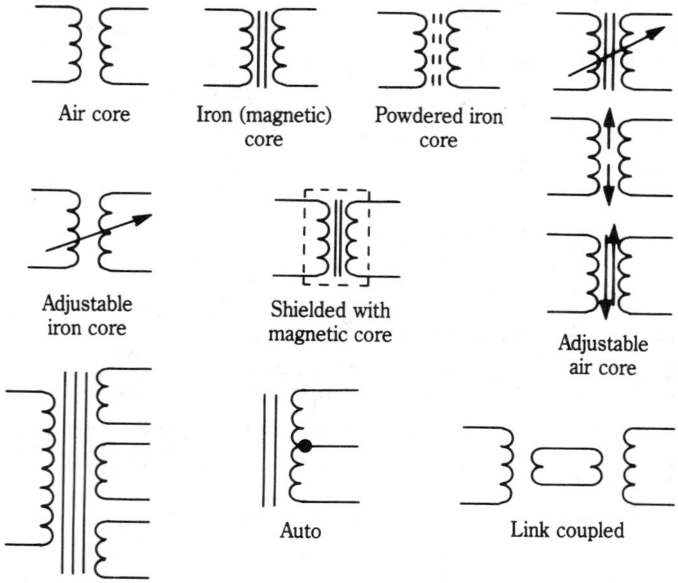

Air core

Iron (magnetic) core

Powdered iron core

Adjustable iron core

Shielded with magnetic core

Adjustable air core

Auto

Link coupled

Power

TRANSISTORS

Bipolar

UJT

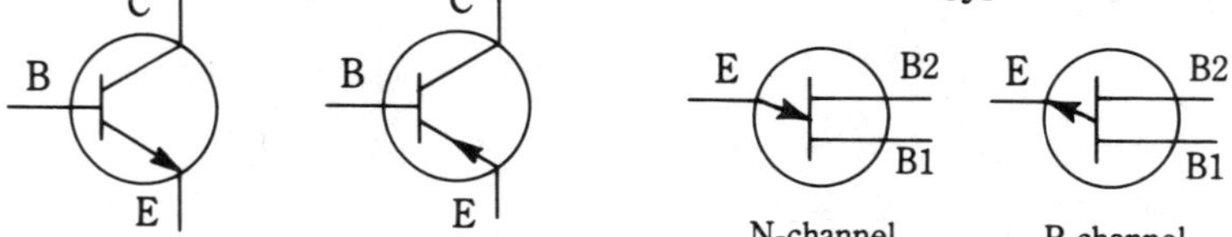

NPN

PNP

N-channel (NP)

P-channel (PN)

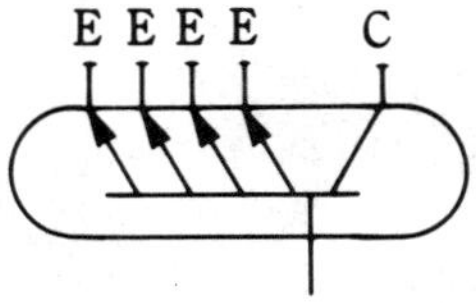

Multiple-emitter (NPN)

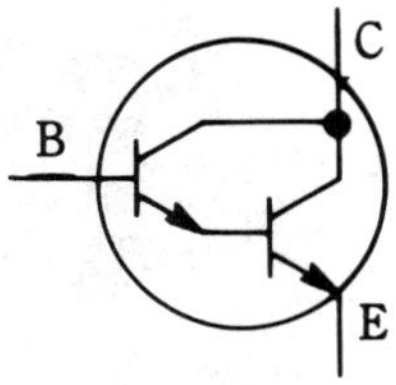

NPN Darlington

Hook (conjugate-emitter connection)

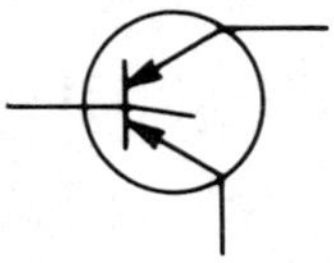

PNPN

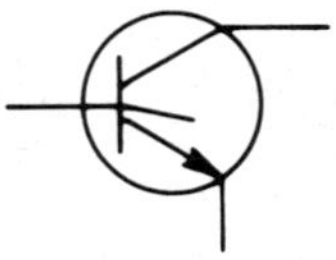

NPN

Tetrode

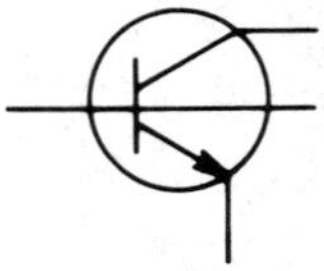

NPN

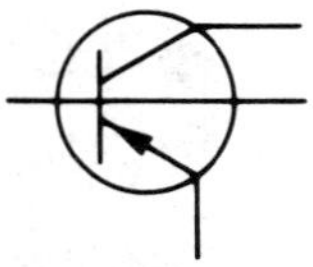

PNP

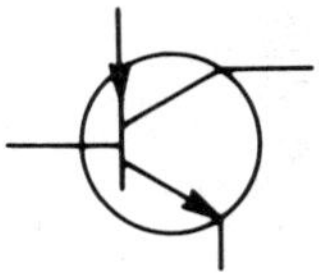

Binistor

PIN triode

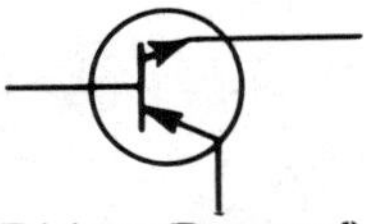

Trigistar (Dynaquad)

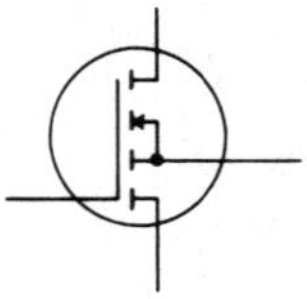

NPN Phototransistor

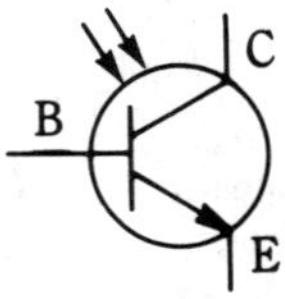

(With base connection)

(Without base connection)

TRANSISTORS—FIELD EFFECT, N-CHANNEL
(for P-CHANNEL, reverse arrow direction)

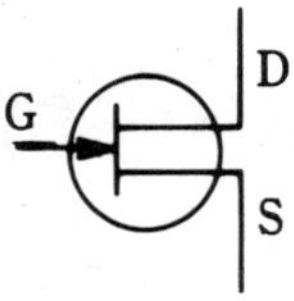

N-channel

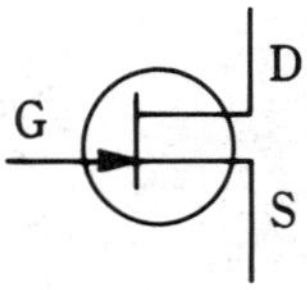

Junction (JFET)

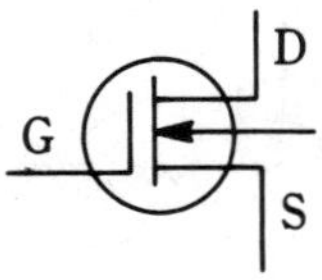

Three-terminal depletion (IGFET)

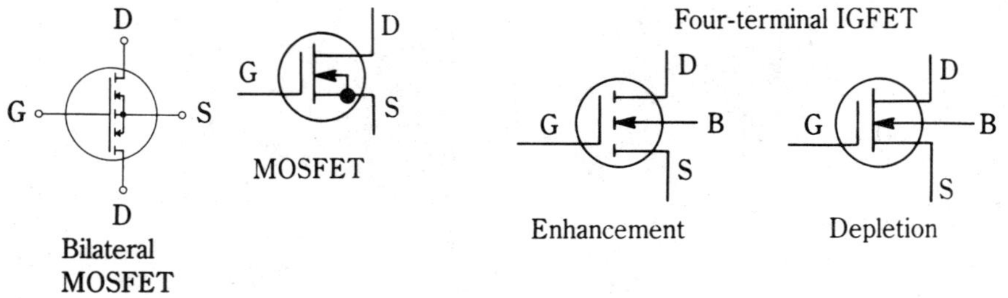

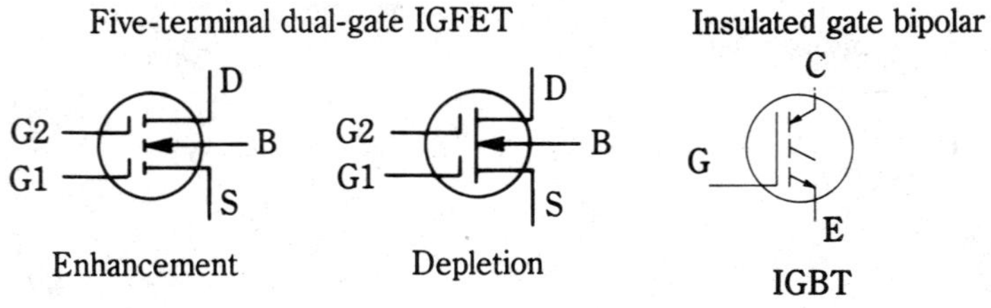

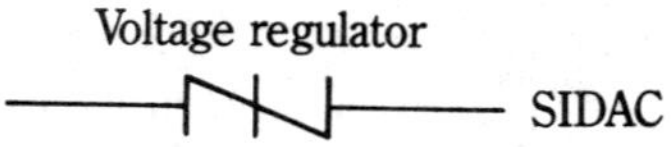

TUBE ELEMENTS

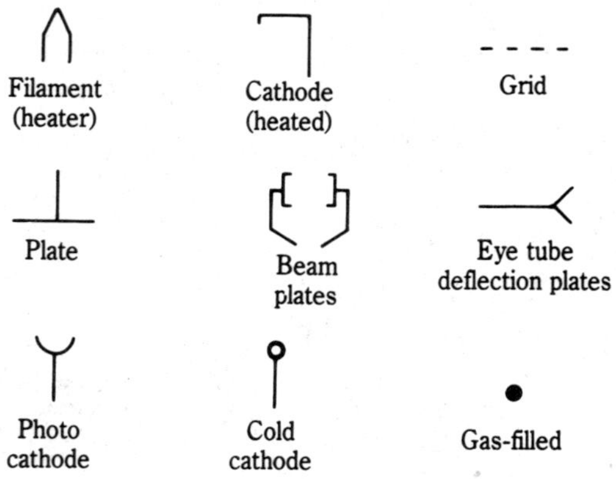

TUBES

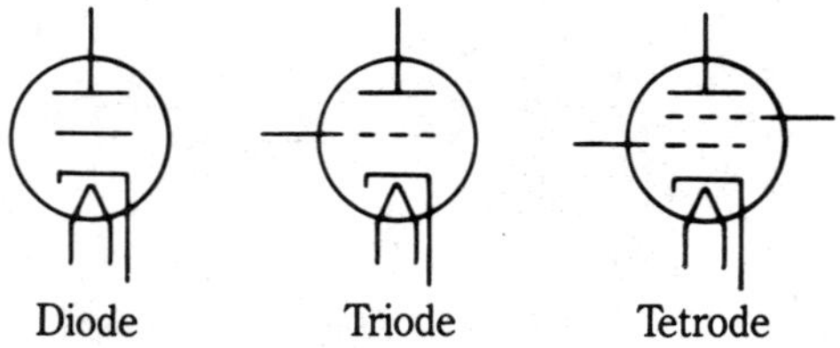

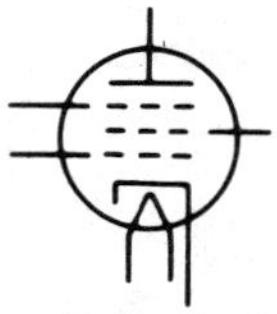
Pentode

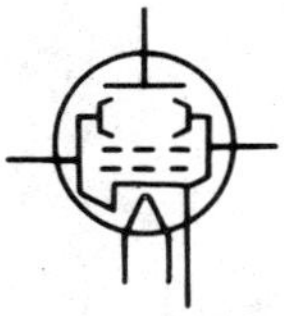
Beam power

Pentagrid converter

Gas-filled rectifier

High-voltage rectifier

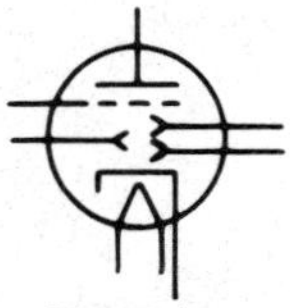
Eye tube

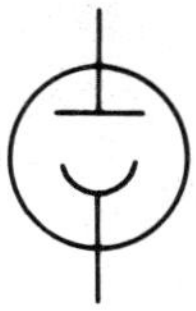
Photo

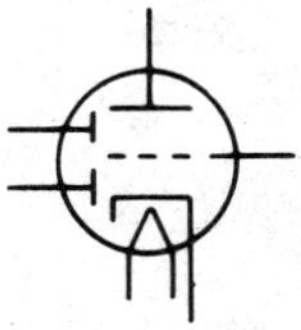
Dual diode triode

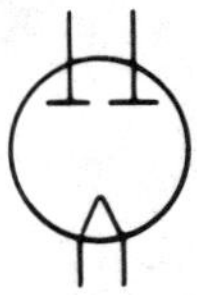
Full-wave rectifier

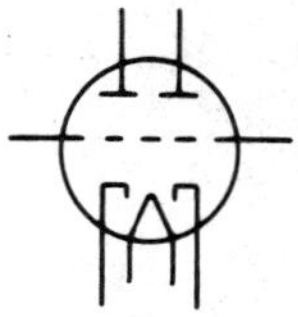
Dual triode

Two-section triode

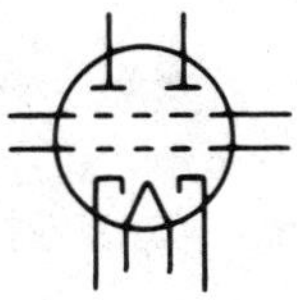
Dual tetrode

Cathode Ray Tubes

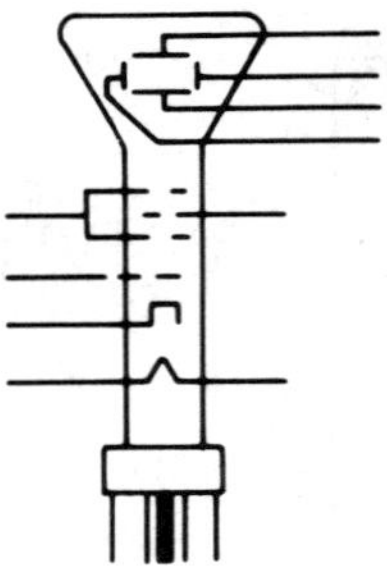

Electrostatic deflection

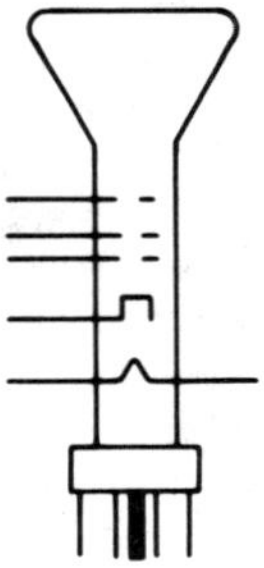

WAVEGUIDES

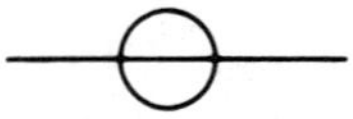

Circular

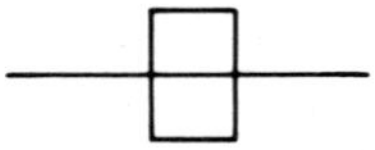

Rectangular

Flexible

Twisted

WIRING AND SHIELDING

Connected

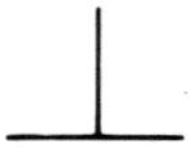

Crossing

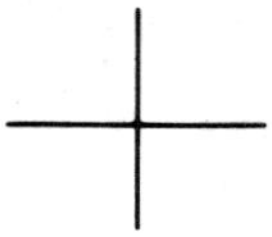

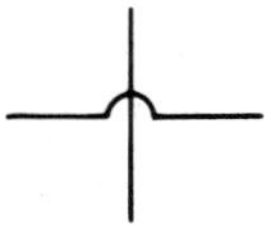

Terminal

Node

Shielding

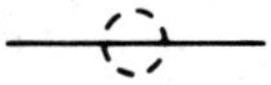

Wire

Multiconductor

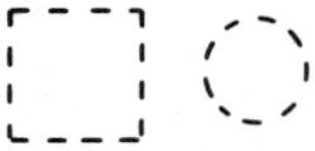

Enclosures

Index